工程翻译：

GONGCHENG FANYI:
BIYI PIAN

笔译篇

邹斯彧 袁 斐 熊 婧◎著

江西高校出版社
JIANGXI UNIVERSITIES AND COLLEGES PRESS

图书在版编目（ＣＩＰ）数据

工程翻译. 笔译篇/邹斯彧,袁斐,熊婧著. --南昌：
江西高校出版社,2023.11（2025.1重印）
ISBN 978 - 7 - 5762 - 4313 - 0

Ⅰ．①工… Ⅱ．①邹… ②袁… ③熊… Ⅲ.
①工程技术—英语—翻译 Ⅳ．①TB

中国国家版本馆 CIP 数据核字（2023）第 213636 号

出 版 发 行	江西高校出版社
社　　　址	江西省南昌市洪都北大道 96 号
总编室电话	（0791）88504319
销 售 电 话	（0791）88522516
网　　　址	www. juacp. com
印　　　刷	三河市京兰印务有限公司
经　　　销	全国新华书店
开　　　本	700mm×1000mm　1/16
印　　　张	14
字　　　数	225 千字
版　　　次	2023 年 11 月第 1 版 2025 年 1 月第 2 次印刷
书　　　号	ISBN 978 - 7 - 5762 - 4313 - 0
定　　　价	68.00 元

赣版权登字 -07 -2023 -812

前　　言

数智时代,随着翻译产业向语言服务产业的转变,语言处理与人工智能、大数据、区块链、云计算等现代技术的深度融合,"语言服务"概念突破传统翻译范畴,被不断赋值和增值,翻译模式不断蜕变革新,内涵和外延不断扩展。

《2019 中国语言服务行业发展报告》将语言服务定义为"以语言能力为核心,以推动跨语言、跨文化交际为目标,向个人或组织提供语际信息转化服务和产品,以及其他相关研究咨询、技术开发、工具应用、资产管理、教育培训等专业化服务的现代化服务业"。我国语言服务行业"已成为中国对外交流的基础设施,成为经济增长的基本保障,成为推进中国'走出去''引进来'的基础性、先导性和支撑性产业"。

国际工程项目是一项以提供工程和技术服务为主的国际官方或民间经济交往活动,建设周期长,涉及面广,需要翻译工作人员全程参与协助。在实际工作中,翻译人员不仅仅提供语言服务,往往还深度参与某个或多个工程工作环节。因而,工程翻译的巨大市场需求量和全新的工作要求为高校在高层次工程翻译人才培养方面提供了全新的认知环境,提出了人才培养方向。

《工程翻译:笔译篇》是根据《翻译硕士专业学位研究生指导性培养方案》的要求编写的一部专业教材,旨在满足翻译硕士专业学位(MTI)研究生、翻译专业本科生高阶学习和工程英语翻译专业人员的学习和工作需求。本教材由七章构成。其中，第一章为工程翻译概述，第二章至第七章分别分析工程项目简介、工程招投标文件、工程合同文本、工程技术文件、工程法律法规和工程学术论著的翻译。

本书得到南昌工程学院"十四五"一流学科"外国语言文学"学科建设经费资助，编写过程中得到于南昌工程学院外国语学院、研究生处和教务处等部门的大力支持以及诸多老师的帮助，在此表示衷心的感谢。

本书中，邹斯彧老师完成 72373 字，袁斐老师完成 91299 字，熊婧老师完成 61328 字。由于时间仓促，本书中或有不当之处，敬请批评、指正！

著者

2023 年 7 月

CONTENTS

目　录

第一章 工程翻译概述

继唐代的第一次翻译高潮之后,明清之际的翻译活动是我国的又一次翻译高潮。明末清初,西方科学技术书籍大规模汉译,形成中国历史上第二次翻译高潮,名为"科技翻译"。根据《欧洲著作汉译书目》记载,在此期间,耶稣会士的汉文西书共 437 部,涉及人文科学、自然科学等各类。其中,自然科学 131 种,包括数学、天文、物理、地质、生物、医学、军事等。这一时期涌现了一大批著名的科技翻译家,如利玛窦、邓玉涵、汤若望、徐光启、李之藻、王徵、李善兰、华蘅芳等。他们的科学译著不仅开阔了中国人的眼界,也因"传播过程的开创性""经典术语的形成"和"广泛的影响力",而为之后的中国人学习现代科学铺平了道路。作为科技翻译的下属类别,近现代工程翻译由此开始。

第一节 工程翻译行业要求

2021 年 11 月 19 日,习近平总书记在出席第三次"一带一路"建设座谈会并发表重要讲话时强调,把基础设施"硬联通"作为重要方向,把规则标准"软联通"作为重要支撑,把同共建国家人民"心联通"作为重要基础。《中国"一带一路"贸易投资发展报告 2022》显示,中国始终将基础设施互联互通作为"一带一路"建设的优先领域,积极对接各方基础设施发展规划,推进一批关系共建国家经济发展、民生改善的合作项目落地,提升交通、信息联通水平,推进能源基础设施绿色转型,夯实高质量共建"一带一路"基础,促进共同发展。基础设施是互联互通的基石,中国企业凭借强大的基建实力,解决共建国家基建堵点,打通"一带一路"建设的血脉经络。

在基础设施方面,中国在共建"一带一路"国家的承包工程新签合同额由 2013 年的 715.7 亿美元升至 2021 年的 1340.4 亿美元,年均增长 8.2%;完成营业额由 2013 年的 654.0 亿美元上升至 896.8 亿美元,年均增长 4.0%。2021 年与共建"一带一路"国家新签对外承包工程项目合同 6257 份。基础设施互联互通是"一带一路"建设的优先领域,是推动共建国家经济发展、改善民生的基石。在如此大规模的工程建设中,工程翻译是项目成功实施的重要因素,对项目的顺利完成起着关键的作用。

一、工程翻译的内涵

翻译包括口译和笔译。笔译又可以分为科技翻译、文学翻译、政论文翻译和应用文翻译等。翻译就是"把一种语言文字的意义用另一种语言文字表达出来"。因此,翻译本身并不是一门独立的创造性科学。它是用语言表达的一门艺术,是科学性的再创作。工程翻译属于科技翻译的一种,涉及工程项目的方方面面,包括各个领域,如土木工程、建筑工程、水利工程、电气工程、计算机工程等。如今,工程翻译是应用性最广的科技翻译之一。

总体来说,工程翻译专业性强——有大量的专业术语,综合性强——涉及工程技术、国际商贸、法律法规等内容,文体类型丰富——包括工程项目简介、工程招投标文件、工程合同文本、工程技术文件、工程法律法规以及工程学术论文等。

二、工程英语的语言特点

工程类文本与其他文体相比,有许多独特之处。工程类文本与专业知识紧密联系,除了包含一些数据(data)、公式(formula)、符号(symbol)、图表(diagram and chart)和程序(procedure)外,在语言、语法、修辞、词汇、体裁等方面都有独特之处。工程类文本的总体特征是语言规范、文体质朴、语气正式、陈述客观、逻辑性和专业性强。其语言特点具体表现为:

1. 词语特点

(1)普通词语专业化

工程英语中,部分通用词语基本上源自英语中的普通词语,这些普通词语被引用到某一专业领域中,虽被赋予新义,但仍与基本词义有紧密的联系,即普通词语专业化。这类词语的特点是一词多义,既有非技术的含义,也有专业含义。例如:pig 在材料工程中意为"金属锭块";cat 在机械工程中意为"吊锚、履带拖拉机";cock 在机械工程中意为"旋塞、吊车"等。

（2）大量吸收外来词语，派生词多

工程英语词语大量借用其他语言表达的情况非常普遍。这类词语的概念大多比较统一，翻译时比较容易处理。例如：thesis（论文）来自希腊语；parameter（参数）来自拉丁语。此外，工程英语派生词的前后缀大多含有拉丁语、希腊语和法语的词根、词缀，例如"inter-"（相互，在……之间）、"micro-"（微小的）、"-scope"（探测仪器）、"-meter"（计量仪器）等。

（3）缩略词多

缩略词书写方便、简洁、容易识别和记忆。在英语科技语中，有大量的词语缩写和缩略形式。缩略词的出现方便了印刷、书写、速记等，但也同时增加了阅读和理解的困难。

2.句子特点

（1）多名词化结构

为使行文简洁、表达客观、内容确切，工程英语中多用表示动作或状态的抽象名词或起名词作用的动词 ing 形式及名词短语结构。

（2）多长句和逻辑关联词

工程英语中虽然大量使用名词化词语、名词短语结构以及悬垂结构来压缩句子长度，但是为了充分说明事理，也常常使用一些含有许多短语和分句的长句，同时还常常使用许多逻辑关联词，以使行文逻辑关系清楚、层次条理分明。

（3）多用一般现在时和完成时

这两种时态之所以在工程英语中很常见，是因为前者可以较好地表现文字内容的无时间性，说明文章中的科学定义、定理、公式不受时间限制，任何时候都成立；后者则可以用来表述已经发现或获得的成果。

（4）广泛使用被动语态

英语中的被动语态要比汉语中的多，在各种文体中都是这样，在工程英语中尤为突出。工程英语的语旨是阐述客观事物的本质特征，描述其发生、发展及变化过程，表述客观事物间的联系。被动语态因此得以大量使用。此外，被动语态所具有的叙述客观性，也使得作者的论述更具科学性，从而避免论述过于主观。

3.语篇特点

（1）语义衔接严密

人称照应和指示照应相对较少,比较照应相对更多,更好地体现了人认识客观事物的过程。高频使用各种逻辑连接手段,体现工程英语语篇逻辑严谨、论证严密等特点。常用同词重述及上义词和下义词来体现条理性,同时避免字面歧义。

(2)少用第一人称和第二人称

工程英语中为避免主观性,少用第一、第二人称,即便是特殊情况下非用不可,也常常使用它们的复数形式以增强论述的客观性。

(3)语法结构严谨

工程英语语篇的主体通常是客观事物或自然现象,旨在阐述客观事物的本质特征及事物之间的联系,描述其发生、发展及变化的过程,语法结构大多比较完整,逻辑连贯且表述畅达,以体现与工程技术相关的科学定义、原理解说及图表说明等诸多表达的客观性、准确性和永恒性。

三、工程翻译译者素养

工程翻译是把技术资料从源文本翻译成目标文本的过程,并且不丢失原本的内容、结构和特点。这就意味着在翻译时,译者不仅要关注语言问题,还要考虑目标地区的文化和规范的细微差别(例如法律规定),以避免产生误解和潜在的冲突。

想要翻译好工程类文档,译者不仅要有深厚的语言功底,还要对翻译内容所属的行业和门类进行深入的了解,对相关的专业术语及基本工作原理要弄清楚,这样才能翻译得专业、准确和规范。另外,这些行业发展日新月异,知识更新速度极快,译者也要随时掌握最新动态,才能更好地完成工程类英语的翻译任务。

1.语言素养

翻译涉及两种语言的转换,对译者的语言素养要求较高。语言素养包括对源语言的理解和对目标语的运用能力。译者的双语能力不仅仅是指通晓基本语言知识,更重要的是指运用语言知识的能力。此外,语言与文化是不可分离的,一个合格的译者应该掌握甚至精通两种语言文化,做好语言和文化的传播者和沟通者。

工程英语翻译者的语言素养还包括译者在翻译工程中要具有一定的语言

敏感性。语言能力不足的译者易被原文复杂的文字和句式束缚和迷惑,翻译出来的译文拗口,可读性差,给读者带来阅读困难。相反,语言能力较强的译者能够辨识语句的不通顺、不合逻辑之处,快速找出译文中可能存在的问题。此时,即使译者由于科技知识的限制,尚不能确定正确的译法,语言的敏感性也能让他发现问题、提出问题。

2. 专业素养

工程翻译所涉及的专业知识比较多,包括机械、电子、计算机、通信、材料等,译者除了需要具备较高的双语能力,还应具备扎实的翻译理论知识与技能,熟悉相关的工程专业背景知识。翻译的过程不仅是语言运用和语义转换的过程,更是逻辑思考和分析的过程。在每一次的翻译实践中,译者都应对文本中涉及的相关专业知识加以归纳和总结,把庞杂的专业知识一点一滴地积累起来。

3. 职业素养

就整体语言特点而言,工程文件的翻译要求简洁严谨。就翻译原则而言,工程行业本身的文体特点决定了译文对"雅"的要求不高,但要绝对遵从"信"与"达",要将这两个要求放在首位,避免使用华丽的辞藻。在翻译时应尽量做到语言严谨、行文简练、逻辑严密,避免使用一些容易产生歧义甚至错误的词句,因为一个细小的翻译错误都可能给公司或客户带来巨大的损失。具体来说,专业译者应顺应工程翻译行业的以下几点要求:

(1)良好的沟通能力

沟通能力是翻译人员必须具备的基本素质。工程项目涉及的内容有很多,不同的国家和地区有不同的语言文化,翻译人员在进行翻译时需要与相关人员进行充分沟通,以确保翻译的准确性和通俗性。因此,工程项目翻译人员必须具备良好的沟通能力,包括语言能力、逻辑思维能力和组织协调能力。

(2)工作细心认真

工程翻译人员在翻译过程中需要保持高度的注意力,时刻保持清醒的头脑,对于原文的内容和结构特点要有充分的理解,并能够随时根据自己的理解进行适当的调整,在翻译时要尽可能地做到准确、严谨、不遗漏。

由于工程翻译涉及很多专业词语,并且各个领域之间也存在一定的差别,因此翻译人员需要做好充分的知识储备,否则会导致翻译工作出现偏差。而如果没有足够的耐心，不够细心，翻译人员也很难在规定时间内完成高质量的翻译工作。

另外，翻译人员还需要对工程文件进行有效的管理，以确保工程文件的完整性、准确性和规范性，其中最重要的是确保文件所包含的信息是最新的、准确的、有效的。当涉及工程文件时，一定要注意文件的完整性。

工程翻译过程中会接触到大量的专业术语、技术标准、法律法规等内容，因此翻译人员必须严格按照相关法律法规和标准规定进行翻译。语言应准确、规范，不能随意改变原文的意思和表述方式。另外，翻译人员还要遵守保密要求，保护客户的商业机密等。

（3）灵活应变

工程翻译工作一般比较繁杂，翻译人员需要随时做好准备。随着技术的更新迭代，工程项目相关知识也在不断更新，技术术语的使用频率不断提高。这要求翻译人员具备很强的应变能力，不能墨守成规。如果不做好准备，翻译人员就可能会在工作中出现问题。

4.道德素养

翻译人员不仅仅是译者，更肩负着对外传播中国文化、讲好中国故事的重任。因此，翻译工作者首先应具备较高的思想政治素质，热爱祖国，忠诚于国家，立场鲜明，具有高度的思想政治觉悟和敏感度。在当今全球经济高速发展的时代背景下，中国提出文化"走出去"的发展战略，加大对外传播力度，力求塑造有国际责任感的大国形象，翻译工作者面临着前所未有的重大使命和责任，因此，思想道德素养的作用举足轻重。

总的来说，工程翻译对译者提出了知识、能力、素养三个方面的要求：熟悉中英工程文本语言和文体特征，掌握工程翻译理论和技巧，了解水利、电力等工程专业知识；具备思辨能力、自主学习能力、跨文化交际能力和综合工程翻译能力（含信息技术运用和项目管理能力等）；具备忠诚担当、科学创新的水利精神以及诚实守信等译商和译德。

四、工程翻译辅助技术

在大数据、人工智能和移动互联网技术的驱动下，语言服务技术正朝着信息化、专业化、网络化、云端化趋势快速发展。这对语言服务人才的技术能力要

求越来越高,掌握现代翻译技术和工具成为时代对译者的必然要求。译者的翻译辅助技术能力主要由以下五个要素构成:

1. 计算机技能

计算机技术的基本应用能力已成为现代翻译职业人才的必备素质。在现代化的翻译项目中,翻译之前需要进行复杂文本的格式转换(如扫描文件转成Word)、可译资源抽取(如抽取 XML 中的文本)、术语提取、语料处理(如利用宏清除噪音)等;在翻译过程中需要了解 CAT 工具中标记(tag)的意义,掌握常见的网页代码,甚至要学会运用 Perl、Python 等语言批处理文档等;翻译之后通常需要对文档进行编译、排版和测试等。可见,计算机相关知识与技能的高低直接影响翻译进度和翻译质量。

2. 信息检索能力

在信息化时代,知识正在以几何级数增长,新的翻译领域和专业术语层出不穷,再聪明的大脑也难以存储海量的专业知识。因此,译者必须具备良好的信息检索、辨析、整合和重构的能力,这也是信息化时代人们应该具备的基本能力。如何在有限时间内从浩如烟海的互联网资源中找到急需的信息? 如何通过专业语料库验证译文的准确性? 这些都需要译者借助信息检索能力才能实现。当代译者应熟练掌握主流搜索引擎和语料库的特点、诱导词的选择、检索语法的使用等,以提高检索速度和检索结果的质量。

3. CAT 工具运用能力

在信息化时代,翻译工作不仅量大,形式各异,且突发任务多,时效性强,内容偏重商业实践,要求使用现代化的 CAT 工具。据 2007 年发布的首个以翻译员为主要关注群体的调查报告——《中国地区译员生存状况调查报告》(传神翻译,2007)的统计,61% 的译者在使用辅助翻译工具,80% 的译者使用在线辅助参考工具。根据《2012 年全球自由译者报告》的统计,65.3% 的受访者认为CAT 工具的使用帮助他们提高了翻译效率。Jared(2013)对 Proz 网站的职业译者做了一项调查,调查显示,88% 受访者使用至少一种 CAT 工具。根据统计,当前各大语言服务公司对翻译人员的招聘要求中,除了语言能力要求,大多数强调熟练使用 CAT 或本地化工具。由此可见,翻译职业对译者的 CAT 工具应用能力要求很高。

4.术语能力

译者术语能力,即译者能够从事术语工作,具有利用术语学理论与术语工具解决翻译工作中的术语问题所需的知识与技能,具有复合性、实践性强的特点,贯穿于整个翻译流程中,是翻译工作者不可或缺的一项职业能力(王少爽,2013)。术语管理已成为语言服务中必不可少的环节,术语管理能力是译者术语能力的核心构成。译者通过术语管理系统可以管理和维护翻译数据库,提高协作翻译的质量,加快翻译速度,促进术语信息和知识的共享,传承翻译项目资产等。因此,当代译者需具备系统化收集、描述、处理、记录、存储、呈现与查询等术语管理能力。

5.译后编辑能力

机器翻译在当前的语言服务行业中具有强大的应用潜力,呈现出与翻译记忆软件融合发展的态势,几乎所有主流的 CAT 工具都可以加载机器翻译引擎。智能化的机译系统可以帮助译者从繁重的文字转换过程中解放出来,译者的工作模式从纯粹手工翻译逐渐转变为译后编辑模式。2010 年,翻译自动化用户协会(TAUS)对全球语言服务供应商的专题调研表明,49.3% 的供应商经常提供译后编辑服务,24.1% 的供应商拥有经过特殊培训的译后编辑人员(TAUS,2010)。因此,译后编辑能力将成为译者必备的职业能力之一。当代译者需要掌握译后编辑的基本规则、策略、方法、流程、工具等。

第二节　工程翻译项目流程

随着全球化的加快,我国的翻译行业也面临着前所未见的挑战。在信息爆炸的背景之下,各类信息的时效性变得极为重要,信息呈现井喷式的增长,现代化的翻译工作也因此发生了翻天覆地的变化:逐渐由传统的个人翻译转向规模大、周期短、专业性强的项目翻译。根据美国项目管理协会对项目的定义:项目是为创造独特的产品、服务或成果而进行的临时性工作。翻译工作是在约定期限内,为满足客户需求而创造出的独特知识产品。因此,翻译工作属于项目范畴。只有将企业的管理经营模式和相关的流程管理运用到翻译项目中,才能够将译员的能力发挥到极致,更重要的是可以提高翻译效率。

工程翻译亦是如此。在面对较大型的工程翻译项目时,只有形成内部协作与外部支持的项目管理体系,规范工程翻译流程和职能分配,才能最大限度地保证项目的质量和速度。

一、工程翻译项目阶段

工程翻译是一个产业链很广的领域,不仅对译员的外语表达能力和知识面有所要求,对翻译公司的项目管理能力也提出了很高的要求。完备的翻译流程应充分涵盖启动、译前、译中、译后各阶段,同时应保证客户、项目经理、译员和审校员的充分互动。在翻译过程中还应将先进的计算机管理技术与翻译实践紧密地结合起来。

1. 启动阶段

首先,项目经理在与客户沟通后,对其需求进行分析并取得工程翻译文件。其次,项目经理根据现有资源,与客户协商项目完成期限、交付形式并确定翻译费用,签订具有法律效力的翻译合同。在项目启动阶段,项目经理要根据实际情况确定时间节点,制订具体的计划并将计划交给客户,双方就计划进行沟通。现在,市场上寻求翻译服务的客户中,有许多客户对译文质量以及自身的翻译需求并不十分明确。这就要求译者恪守职业道德,尽量为客户提供高质量的译文。

2. 译前阶段

译前阶段并不属于传统意义上的文字翻译过程,但却是工程翻译项目必需的阶段。首先,项目管理者要依据翻译原件的难易程度、译件用途,以及各译员所擅长的领域及翻译速度来分配任务。其次,项目团队将客户所交付的格式复杂的文档转换成便于翻译的文件格式,也可以运用计算机辅助翻译软件提取项目中出现频率较高的专业术语对其进行预翻译并制作术语库,以便在多人合作翻译的时候保证术语统一。在译前进行术语统一,可避免出现后期术语译法不一致的现象,减少译后审校的压力,最大限度地提升翻译效率。在译前阶段,为了确保项目成员有效沟通,还可以提前选好一款协同共享软件让团队成员在线共享翻译记忆库、术语数据库等项目资源。

在译前阶段所做的一切其实都是在为保证优秀的译文质量做准备。想要获取高质量的译文,优秀的项目经理、译员以及审校员就要通力配合。

3. 译中阶段

在译中阶段,翻译团队应根据翻译原件的内容类型和用户的需求统一格式、术语及译件形式,并且做好翻译过程中疑难问题的解决工作。工程翻译文本专业性较强,一旦碰到疑难问题,译者可以查询专业资料,或者咨询专家,确保翻译万无一失。

为了确保翻译项目的顺利运作,监控工作需贯穿整个译中阶段。此时,质量监控不仅仅是对译件质量的监控,还包括对翻译过程的监控。译件质量监控指的是对译文的审校,不仅包括语言表达方面的审校,还包括专业知识方面的审校。语言表达方面的审核内容包括:译文是否有漏译、错译;内容和工程术语是否准确;语法和词法搭配是否准确;是否遵守与客户商定的有关译文质量的协议;译者的注释是否恰当;译文的格式、标点、符号是否正确。尤其要注意工程文本中的数字、单位及术语的准确性,避免给客户造成损失。如果译员不是工程翻译方面的专业译员,那么最好另找专业人士进行专业知识方面的审校。最后的审校环节是对排版细节等进行检查,并给予相应反馈。

对翻译过程的监控主要指要进行有效的时间管理及进度控制,这些对于翻译项目的成败起着举足轻重的作用。良好的时间管理是保证快速响应客户需求、及时交付项目的关键。在多人合作的翻译项目中,进度控制应当获得高度的重视。进度控制不仅涉及译前的准备工作,还应当在翻译过程中有所体现。首先要制作详细的工程翻译项目日程表,使译员和审校员严格按照日程表上的进度要求开展工作,这样可以系统、有效地对整个翻译项目进行统一控制,避免出现进度混乱的局面。日程表包括两方面的内容:一是对整个翻译项目进行细致的分解;二是明确每个节点要完成的任务要求。在实施过程中,项目经理应该督促译员进行有效的沟通,同时严格督促译员按照项目进度表完成各个节点的任务,及时跟进项目进程,并依据项目计划与质量标准对项目偏差及时进行调整。这一环节要求项目经理对翻译项目的内容、各译员及审校员的翻译水平非常了解,这样项目经理才能做出合理的安排。

项目经理作为项目团队与客户间的联系人,需与客户即时沟通以准确获取客户的不同需求。此外,项目团队成员间也需要进行交互式沟通,促进信息的交换,培养信任感,增强团队的凝聚力。为了控制项目费用、保证现金流的健康运行,项目经理还需了解不同译员擅长的工程领域及报价,以确保翻译项目的顺利完成和项目资源配置收益率的最大化。

4.译后阶段

翻译以及审校工作完成以后，整个翻译项目也接近尾声。在此阶段，项目经理需要将译稿统一收回，经整体审校确定无误后按照客户要求进行统一排版、打印、装订、印刷等工作，要确保译文在格式、术语、风格等各个方面都保持一致，并且符合客户的要求。项目经理依照客户要求交付翻译产品，接受客户验收和评价，依据客户的反馈，分析问题并提出可行的解决方案。每一次的翻译都是不断完善以满足客户需求的循环推进过程。在翻译项目得到客户的认可之后，项目经理根据翻译合约的各项条款，一一核查、兑现，与客户进行翻译费用的结清工作。

最后对本次翻译任务进行总结——总结经验教训和得失，同时更新最终的翻译记忆库和术语库，为下一次工程翻译项目的实施提供便利。

二、工程翻译项目流程管理

在工程翻译项目中，项目经理需要根据客户的需要，在人员、时间、成本、风险、翻译技术以及质量等方面制订相应的宏观计划，同时和其他翻译项目协调好关系，做好以下几方面的管理工作。

1.人员管理

人员管理指的是对项目人员的选择、团队协调、高效率组织结构规划的管理以及对人员冲突的及时调解。译前需要选择具有相关经验的参与人员建立团队，译中主要进行人员监控和激励。监控的目的在于人员是否按照既定计划开展工作，一旦出现问题及时解决。与此同时，一些大型项目耗时长，问题复杂，容易给人疲惫感，所以激励措施至关重要，可以通过定期团建、举行交流会来增强团队的凝聚力。翻译工作结束后，团队可以就整个项目的完成情况进行总结、交流。

2.时间管理

时间管理指的是为完成高质量项目，根据交付时间而采取的具体行动。项目经理首先需要根据项目的复杂程度对不同项目的交付时间进行合理安排，然后针对每个项目估算活动资源，即所需要的材料、人员、设备等，并且对每项分任务确定好时间节点，最后编制一项完整的进度计划。这是译前的准备工作。

译中主要进行时间监控，监督项目状态以更新项目进展，管理进度基准变更后需要调整时间安排。在调整时间节点时，关键在于计算各类浮动时间，确定哪些活动进度无法修改，并且对灵活性较大的时间安排进行合理压缩。

3. 成本管理

成本管理主要包括成本估算、预算和控制三个过程，旨在确保在合理预算内完成整个工程。项目管理初期，项目经理需要根据客户的需求以及客户所给的预算，进行合理估算并且制订整体预算方案，明确各类可能出现的支出，以确保留有正常的盈利。译中主要是把控风险，尽量按照整体计划进行各类支出的估算。在成本管理中，应注意对其他供应商的管理。当翻译项目较多，译员人数不够时，可以考虑将一些翻译内容进行外包，在不影响质量的情况下尽可能压低成本。成本还涉及技术、设备等非人力资源，对这些资源及时更新从长远看可以提高效率、降低成本。但是，如果设备或者相关技术软件费用过高，且要对工作人员进行培训，那么效益就不一定能保证。这时，项目经理需要仔细计算成本。因此，成本管理尤其重要，涉及多个相关领域。

4. 风险管理

风险管理主要体现在风险管理规划、风险识别、风险分析、风险应对规划这几个方面。在项目规划阶段，项目团队需要针对可能发生的项目风险制定完善的对策，以规避、转移、降低风险。某些不可抗力因素可能造成人员空缺，因此需要提前进行人员储备，以免影响整个进度。例如，发生火灾会造成资源损毁，项目管理人员就需要提前制定备份机制，可以运用"云技术"使资源云端化，降低资料损毁、丢失造成的风险。

三、工程翻译项目流程中的问题及对策

当前国内翻译市场上的工程翻译流程问题还是很突出，主要集中在以下几方面：首先，部分翻译公司未设立完善的翻译流程。翻译流程的不完善反映了翻译公司在项目管理中存在种种漏洞和问题，甚至引发严重的质量事故。其次，翻译流程中过于注重翻译，而忽视了审校，"译"和"审"的重视程度不匹配。再次，翻译各阶段和环节缺乏互动和反馈，工作交流协同程度不够。单一大型文件的翻译往往需要多个译员共同完成。在没有较好地沟通和协作的情况下，对同一部分的内容进行翻译时容易出现前后用词不一致和歧义的问题，无法保

证译文的质量。

针对当前国内翻译市场翻译流程混乱的现状和其他弊病,翻译机构应该多方合作,重新梳理翻译流程,制定行业标准和规范,提高企业和从业人员的自觉性。

第一,重新梳理翻译流程,细分步骤,建立完善翻译流程的指导性框架。完善的翻译流程应充分涵盖译前、译中、译后各阶段和各参与角色,包括前期在译者、稿件、项目、团队方面所做的各项准备和处理工作,并充分同客户进行沟通,了解客户的要求和需要;能够制约和追踪翻译过程中译者的行为和进度,同时涵盖中途的质量控制和技术支持等;应增强各环节之间的互动,译文完成后应进行各项验收、反馈和归档等工作。此外,完善的翻译流程还应体现质量管理工具和电子工具的作用,并配备应急方案,为翻译事故或应急事件提供即时的解决方案。

第二,制定翻译流程相关的标准和规范。国家标准《翻译服务规范 第1部分:笔译》(GB/T 19363.1—2003)对翻译、审核、质量保证、资料存档、顾客反馈和质量跟踪等内容都进行了梳理,涵盖了翻译人员和审校人员的资质、译前准备、审校等方面。将梳理清楚的完善的翻译流程上升到标准或规范的层面,将进一步规范翻译活动各主体和从业人员的行为,保证翻译有标准可依、有规章可循。企业自身也可以制定相关的业务规定或项目指南,在制度层面予以保证。

第三,从操作层面而言,翻译流程起到规范企业或个人翻译行为、防范译文质量风险的作用。因此,完善的翻译流程必须包括多个控制环节,每个环节紧紧相扣,必须有各自的"准入"门槛和"放行"标准。一旦某个步骤或环节出了问题,"瑕疵品"严禁进入下一环节。这样不仅降低了某些环节遗漏可能带来的质量风险,也增加了前端和后端的紧密度,还能够从全局控制译文质量,使翻译项目能够顺利实施。

第三节　工程翻译标准与原则

我国翻译及语言服务行业发展迅猛,产值稳步上升,共建"一带一路"国家的翻译业务量显著增长,行业标准化建设稳步推进,行业规范化管理水平进一步提高。2021 年,全球以语言服务为主营业务的企业总产值预计首次突破 500 亿美元;中国含有语言服务业务的企业达 423,547 家,以语言服务为主营业务的企业有 9656 家,企业总产值实现稳步增长,达到 554.48 亿元,相较 2019 年增长了 25.6%。行业标准化建设有力地推进了翻译行业的标准化和规范化进程。

一、翻译行业标准和规范

2001 年起,国家质量监督检验检疫总局陆续制定并出台系列翻译服务国家标准。这些国家标准的出台填补了翻译领域国家标准的空白,开启了中国翻译服务标准化的进程,推动中国翻译行业向规范化管理迈出了重要的一步,也使我国在国际翻译标准领域处于领先地位。这些标准还是我国首批以中英文同时发布的国家标准。中国翻译协会还根据行业需求,组织制定了多部行业规范,作为国家标准的有益补充,对规范语言服务行业发挥了重要的作用。

2003 年 11 月 27 日,国家质量监督检验检疫总局发布《翻译服务规范　第 1 部分:笔译》(Specification for Translation Service-Part 1: Translation,中华人民共和国国家标准 GB/T 19363.1—2003,2004 年 6 月 1 日开始实施)。该标准修订版于 2008 年 7 月 16 日发布,2008 年 12 月 1 日实施,标准号为 GB/T 19363.1—2008。

《翻译服务规范　第 1 部分:笔译》是我国历史上第一次对翻译行业制定的国家标准,是服务行业的推荐性国标。该标准根据翻译服务工作的具体特点,以 2000 版 GB/T 19000/ISO 9000 质量标准体系为指引,目的在于规范行业行为,提高翻译服务质量,更好地为顾客服务。本标准规定了提供翻译服务的过程及其规范,适用于翻译服务(笔译)业务。本标准首次以国标的形式明确了"翻译服务"的定义及内涵,即"为顾客提供两种以上语言转换服务的有偿经营行为";首次以国标的形式对翻译服务方的业务接洽、追溯性标识、业务管理、质量保证、资料保存、顾客反馈、质量跟踪等方面,提出明确的规范性标准。

2005 年 3 月 24 日,国家质量监督检验检疫总局、中国国家标准化管理委员会发布《翻译服务译文质量要求》(Target Text Quality Requirements for Translation Services,中华人民共和国国家标准 GB/T 19682—2005,2005 年 9 月 1 日开始实施)。《翻译服务译文质量要求》(以下简称《要求》)就译文质量的基本要求、翻译译文中允许的变通、译文质量评定做出规定。《要求》对译文质量提出了基本要求——忠实原文、术语统一、行文通顺,同时提出了译文质量的具体要求,分别就翻译过程中最常见的数字表达、专用名词、计量单位、符号、缩写词、译文编排等提出了处理规范。此外,《要求》就翻译服务译文中常见的需要进行特殊处理和表达的若干问题提出了变通处理办法。该标准提出以译文使用目的作为译文质量评定的基本依据,对译文质量要求、译文质量检验方法制定了规范性标准。

二、工程翻译标准

工程翻译属于科技翻译大类的下属分支。工程类文本具有科技文本的共同特点,同时也具有专业特殊性。总体而言,工程类文本属于信息型文本,具有专业性、信息性、匿名性和匿时性特点。所谓"专业性"指的是工程类文本通常具有典型的专业唯一性,文本内容专业性强,具有行业知识门槛;"信息性"指的是工程类文本的核心功能是真实、准确地传递专业信息;"匿名性"指的是工程类文本通常无署名、无版权;"匿时性"指的是工程类文本通常采用现在时态,不强调过去与未来时态。

工程翻译应遵循工程类文本的特征,形成其特有的翻译标准。所谓"翻译标准",也就是衡量翻译质量的尺度。根据《翻译服务译文质量要求》相关规定,译文应当遵循严复提出的"信达雅"标准,具体表现为:

第一,信息之信,即忠实原文、术语准确、信息无误。

工程翻译应将译文信息的准确性和精确性置于首位。不同于重"美学""情感"的表达型文本和重"感染"的呼唤型文本,工程类文本属于信息型文本。工程翻译应以准确传递信息为主要目的,译文应忠实原文,同时关注信息传递效果,用工程术语表达工程专业内容,避免信息错误、语义模糊和文辞多义等情况,内容不宜过分"增""删""减""改"。译者应明确术语概念范畴和行业限制。在特定的专业语境下,术语应具有一元单义性、表述精准性和语义稳定性。所谓"一元单义性",指的是术语往往由上下文语境或专业行业定义,其所处文本环境决定其意义。所谓"表述精准性",指的是大部分术语表述有其定式,些微的更改或许会影响其语义或正确性。所谓"语义稳定性",指的是术语之义相对而言较为固定,语篇等因素对其内涵的影响较小。

不同文本或不同专业行业语境下,同一词语的意义也有所区别,如:tender design(招标设计)一词中,"tender"应与日常使用的语义"温柔的、敏感的、娇嫩的"进行区别;overflow earth-rock dam(溢流土石坝)一词中,"overflow"的译文应依据专业性和行业背景,区别于日常用语"溢出";shaft intake(竖井式进水口)一词中的"intake"应根据上下文补充完整信息,而非简单译作"摄取,吸入";rigid pipe(刚性管)中的"rigid"一词,若译为"严格的,僵化的"则与原文差之千里。

第二,言语之达,即术语统一、行文通顺、表述规范。

翻译活动通常涉及多位译员和多个环节。在整个翻译过程中,翻译团队须做到术语翻译译文上下统一。工程英语语法结构严谨,多用被动语态和it等特殊句式,时态、人称、结构变化性不强,程式化明显。译文也应当保证修辞正确、逻辑合理、文理通顺。

【例1】原文:Electrical energy can be stored in two metal plates separated by an insulating medium. Such a device is called a condenser, and its ability to store electrical energy is termed capacitance.

译文:电能可以储存在被一绝缘介质隔开的两块金属板中。这样的装置称为电容器,它储存电能的能力就称为电容。

【例2】原文:It has been proved that induced voltage causes a current to flow in opposition to the force producing it.

译文:已经证明,感应电压使电流的方向与产生电流的磁场力方向相反。

在上述案例中,为实现译文语句通顺,原文中的被动句转换为主动语态或无主语等句式,以顺应中文少见"被"字句的语言特征。

第三,风格之雅,即语言专业、陈述客观、逻辑严密、结构整齐。

译文应符合工程类相关工作的语境三要素——语场(field of discourse)、语旨(tenor of discourse)和语式(mode of discourse)的特征与要求。语场指的是文本题材或话语范围;语旨指的是讲话者(作者与读者)之间的角色关系;语式指的是口头语或书面语、自然语或人工语等交际媒介等。三要素之中任何要素发生变化,都势必导致语言变异,由此产生不同的语域。

工程类文本种类众多,涉及不同讲话者、角色关系、交际媒介等语域因子。总体而言,工程类文本在风格上,陈述客观,逻辑严密,表述简洁,言语专业,结构完整、对称,译文风格也当如此,以体现工程类文本的语言美和形式美。例如:

原文:The administrative staff for the project will be composed primarily of Chinese engineers and senior technicians. It is proposed that the skilled, semi-skilled and unskilled workers will be employed locally. Local manpower required for the works shall be planned at the time of commencement in the light of meteorological condition, construction progress and the same shall be submitted to the engineer for approval. The initial proposed manpower is tabulated below.

译文:项目的管理人员主要以中国工程师和高级技师为主。建议技工、半熟练工和普工在当地招聘。项目需要的当地人工招聘应在开工时根据气象条件和工程进度来做计划,而且同样需提交给监理审核。初步的人力计划如下表。

三、工程翻译原则

习近平总书记在中共中央政治局第三十次集体学习时强调:“讲好中国故事,传播好中国声音,展示真实、立体、全面的中国,是加强我国国际传播能力建设的重要任务。”二十大报告围绕“增强中华文明传播力影响力”,提出“加强国际传播能力建设,全面提升国际传播效能,形成同我国综合国力和国际地位相匹配的国际话语权”。

当代语言服务者应担负起传播中国声音、提升国际传播效能的任务,工程翻译译员应根据工程类文本的专业性、信息性、匿名性和匿时性等特征,优化自身专业知识结构,提高快速学习能力,遵从以下工程翻译原则:

(一)专业原则

工程翻译应注重词义之专业、句式之专业和表述之专业。

翻译过程中,上下文的重要性远远大过词典。工程翻译中,专业语境决定了术语翻译中的词义选择。词语的翻译与专业和行业背景有着紧密的内在联系。

显性的专业词语具有典型的单一性,词义明确,但隐性的专业词语模糊性较强,往往在日常生活和多个专业行业中出现,易造成误译。因此,翻译者须紧密联系上下文,准确判断专业背景,确定语义。

（二）精准原则

工程翻译中的精准原则指的是信息传递准确,包括语言语义内涵与外延的准确移植以及数据的精准传递。

工程类文本通常涉及大量数据,工程翻译不仅仅要保证语言文字信息传达准确,数据的精准也尤为重要。数据或词语的误译往往导致重大经济损失或法律责任。如出现误译,译员须承担一定的法律或经济责任。因此,精准原则极为重要。

（三）美学原则

工程翻译须遵循特定文本的风格特征和美学要求,形成特有的语言简洁、陈述客观、结构整齐、逻辑严密的美学特征。

翻译既是科学,也是艺术。工程翻译之"美"不同于文学艺术作品的绚丽之美,表现出科学严谨之美、整齐有序之美、精确和谐之美。

（四）价值原则

"科学无国界,科学家有祖国。"语言服务者应当始终清楚地意识到自身所肩负的"翻译中国"之职责,具备忠诚担当、诚实守信、科学创新的工程人精神和家国情怀等译商和译德,在翻译活动中应坚守端正态度、忠实传译、保持中立、保守秘密、遵守契约、合作互助、妥用技术和提升自我等译员职业道德。

第四节　工程翻译的机遇与挑战

从改革开放到加入世界贸易组织再到"一带一路"建设,我国涉外工程项目逐年增加。自2010年以来,我国对外承包工程营业额连续3年超过千亿美元。2019年,美国《工程新闻纪录》公布的全球最大的250家承包商中,仅中国内地就有75家。其中,3家进入前10强,10家进入前50强。快速增长的涉外工程项目带动了工程翻译的发展和高层次工程翻译人才的需求。翻译人员在工程建设的各个环节参与度较高,凭借语言优势与当地人员建立了高效便捷的沟通途径。翻译人员作为"工具"和"桥梁"的时代早已结束,他们不仅仅提供语言服务,往往直接参与某些工程项目的某些工作环节。此外,数智时代翻译模式的变化和疫情时代的结束,使工程翻译面临新的机遇与挑战。

一、行业机遇

根据中国翻译协会公布的《中国翻译及语言服务行业发展报告 2022》。2017—2021 年,我国翻译及语言服务行业发展迅猛,产值稳步上升。2021 年,全球以语言服务为主营业务的企业总产值预计首次突破 500 亿美元;中国含有语言服务业务的企业 423,547 家,以语言服务为主营业务的企业有 9656 家,企业总产值实现稳步增长,达到 554.48 亿元,相较 2019 年增长了 25.6%。共建"一带一路"国家的翻译业务量显著增长;阿拉伯语、俄语、德语、英语和白俄罗斯语为市场急需的五个语种。随着全球疫情的结束,翻译行业产值预计进一步提高。

人工智能技术不断创新,机器翻译在行业中的应用越来越广泛;"机器翻译 + 译后编辑"的服务模式得到市场普遍认同,其最大优势是提高翻译效率、改善翻译质量和降低翻译成本。2021 年,开展机器翻译与人工智能业务的企业达 252 家。91% 的语言服务企业认为,采用"机器翻译 + 译后编辑"模式提高了翻译工作效率。

93.8% 的语言服务需求方表示,最近两年在共建"一带一路"国家有频繁的投资或贸易活动;81.7% 的语言服务企业表示,共建"一带一路"国家的翻译业务量呈现增长态势。由此可见,工程翻译有着巨大的市场潜力。

二、行业挑战

工程翻译市场在继续扩张的同时,也面临巨大的挑战。

首先,翻译行业准入制度尚不完备。翻译人员翻译能力参差不齐,专业度不完全合格;部分翻译人员甚至未经过专业学习和训练,不具备翻译资质。

中国外文局负责实施与管理的一项国家级职业资格考试——全国翻译专业资格(水平)考试已被纳入国家职业资格目录清单,是一项在全国实行的、统一的、面向全社会的翻译专业资格认证。设立这一考试的目的是加强翻译行业管理,规范翻译就业市场,促进翻译行业人才队伍建设,科学、客观、公正地评价翻译专业人才水平和能力,使中国翻译行业更好地与国际接轨,为中国与世界各国政治、经济、文化、教育等领域的交流合作提供翻译人才资源。然而,在真实的翻译市场上,大量翻译人员或承担翻译工作的人员并非翻译专业毕业生,而是语言学习者甚至工程相关行业人员。

其次,翻译质量参差不齐。工程翻译质量参差不齐的原因较为复杂,或许是翻译人员未经过专业训练,或许是专业翻译人员翻译能力、知识结构、翻译时长、翻译报酬,甚至职业道德、自我约束力等因素的影响。

再次,翻译人员专业知识欠缺。翻译人员多为语言学习者或者翻译专业毕业生,缺乏必要的专业行业知识,导致在原文的理解、术语语义的确定等方面出现偏差,从而出现误译。让译者进入误区的并不是生僻的词语或显性的专业词语,而是常用词或隐性的专业词语。常用词或隐性专业词语的词义较丰富,含义范围较宽;如不考虑具体语境,难以确定专业或行业背景,译文难免背离原文。

最后,翻译技术助力翻译工作的同时,也对翻译人员的翻译能力提出更高的要求。全球范围内,人工智能技术的发展和应用深刻改变着人类的社会结构。为抢抓人工智能发展的重大战略机遇,加快建设创新型国家和世界科技强国,2017年国务院发布《新一代人工智能发展规划》(以下简称《发展规划》),将发展人工智能作为国家重点发展战略之一。人工智能技术之一的自然语言处理技术以及根据此技术开发的机器翻译技术,成为重点研究领域。《发展规划》指出,自然语言处理技术具体研究短文本的计算与分析技术、跨语言文本挖掘技术、面向机器认知智能的语义理解技术、多媒体信息理解的人机对话系统。

2016年谷歌公司发布神经网络机器翻译(NMT)系统后,全球范围内NMT机器翻译成为主流的机器翻译系统。NMT翻译的译文质量,与SMT(统计机器翻译)相比,有了大幅提高。自此,基于NMT技术的各类机器翻译在全球范围内开始广泛应用,以机器翻译(MT)的译文为基础、进行人工译后编辑(PE)的翻译模式成为语言服务行业的新兴翻译形态。根据美国卡门森斯咨询公司的全球语言服务市场报告,机器翻译的译后编辑成为全球语言服务企业占比第三的服务收入来源。

翻译技术在语言服务行业应用越来越广。根据中国翻译协会发布的中国语言服务行业发展报告,计算机辅助翻译、翻译质量保证工具、搜索引擎和桌面搜索、翻译管理系统、机器翻译及译后编辑成为语言服务需求方应用排名前五的翻译技术；翻译管理工具、搜索引擎和桌面搜索、计算机辅助翻译、术语管理工具、翻译交易平台成为语言服务提供方应用排名前五的翻译技术。

　　2004 年设立的欧盟翻译硕士（EMT）教学内容具有较大的参考价值。欧盟翻译硕士发布的欧盟翻译硕士能力框架（EMT Competence Framework 2017）报告，将技术作为五项翻译能力之一（五项能力分别是语言与文化能力、翻译能力、技术能力、个人与人际关系能力、服务提供能力）。欧盟翻译硕士能力框架中的技术包括办公软件、搜索引擎、语料库工具、预处理工具、机器翻译系统、翻译工作流程管理工具等。EMT 能力框架将"技术能力"作为五项翻译能力之一，深刻表明翻译技术在翻译专业人才培养中的重要地位。

第二章　工程项目简介

工程项目简介是一个工程项目顺利开展的重要文件，主要是指对一个工程项目进行总体的描述。内容包括工程名称、规模、性质、用途、资金来源、投资额、开竣工日期、建设单位、设计单位、监理单位、施工单位、工程地点、工程总造价、施工条件、建筑面积、结构形式、图纸设计完成情况、承包合同等。

第一节　背景分析

一、工程项目简介的作用

1. 为项目管理提供基础数据。工程概况可以为项目管理提供基础数据，如项目的目标、范围、进度、预算等。这些数据可以作为项目管理的基础，为项目的进展提供依据。

2. 为项目决策提供依据。工程概况可以提供项目的风险、资源等方面的信息，这些信息可以作为项目决策的依据，帮助项目管理者制定合理的决策。

3. 为项目沟通提供依据。工程概况可以为项目的参与者提供一个共同的理解和认识，帮助各方进行有效的沟通和协作。

4. 为项目评估提供依据。工程概况可以作为项目评估的依据，帮助评估人员对项目的成果、效益等方面进行评估。

二、工程项目简介的内容

根据对象的不同，工程项目简介内容的侧重点也会有所不同。用于招投标目的的工程项目简介的内容必须包含：建设内容、建设规模、投资总额等，项目的主要内容、创新点、技术水平及应用范围，简述项目的社会经济意义、现有工作基础、申请项目的必要性。用于项目施工管理的工程项目简介还需要说明资金来源、投资额、开竣工日期、建设单位、设计单位、监理单位、施工单位等信息。而用于项目建成后的评估展示，则重点突出其效益运行情况。

然而,无论出于什么目的、针对什么对象,工程项目简介都必须包含以下几项内容:

1. 项目名称。工程项目名称是其最基本的标识,通常由项目的类型、地点、规模等因素组成。

2. 项目背景。工程项目的背景是指项目实施的原因和必要性,通常涉及政策、技术、市场、环保等方面。

3. 项目规模。工程项目的规模是指项目所涉及的空间范围和工程量的大小,通常以建筑面积、容积率、总投资等指标来衡量。

4. 项目目标。工程项目的目标是指项目实施后所要达到的效果和目标,通常包括经济效益、社会效益、环境效益等方面。

5. 项目实施计划。工程项目的实施计划是指项目实施的时间、步骤、资源配置等方面的规划,通常包括前期调研、设计、施工、验收等阶段。

6. 项目组织架构。工程项目的组织架构是指项目实施所需要的组织机构和人员配置,通常包括项目部、设计单位、施工单位、监理单位等。

7. 项目风险评估。工程项目的风险评估是指对项目实施过程中可能出现的各种风险进行评估和预测,以便采取相应的措施进行风险控制。

三、译前准备

在翻译任务开始时,译者首先要分析源文本的语言特征,并根据语言特征和翻译要求进行译前准备。

从词语水平上看,工程项目简介源文本中包含许多具有较高专业价值的专业词语,如下表中的词语:

鱼道	fishway
坝轴线	dam axis
门库坝	gate chamber dam
应力	stress
应变	strain
闸墩	pier
坝肩	dam abutment
径流式	run-of-river

续表

枯水围堰	starter cofferdam
坝顶高程	crest elevation

译者应该查阅专业词典或现有的水利工程语料库,对已有的术语进行整理,为后续工作做准备。从句子结构层面看,原文的句子结构主要由并列分句和祈使句组成,所有的句子都有严格的逻辑关系。对于句子和语篇的翻译,译者应参考现有的工程项目介绍文本及其翻译版本的语言特点,采用不同的翻译手法,力求满足翻译要求。

四、翻译要求

第一,译文应忠实地表达原文的信息。

工程类文本专业性强,对信息的准确性要求较高。译者应该花大量的时间对专业术语进行分析,然后选择合适的词语完整地表达其语义。此外,工程项目简介文本具有高度的逻辑性,译者需要分析不同的句子结构,以提高翻译质量。因此,译者不仅要保证翻译的准确性,还要对水利工程专业知识有深刻的理解。

第二,译文应流畅易读。

由于汉英语言的差异,很多句子不能逐字翻译,否则翻译出来的文本会很死板,读者阅读起来也很困难。因此,为了使目的语读者有效理解文本的意思,译者需要采用不同的翻译方法,实现源语与目的语的等值。

第三,译语应符合工程项目简介文本的语言特点。

准确的术语、简洁的文字、逻辑严谨的表达、客观的描述,可以使翻译更符合工程类文本的语言特点,确保译文读者和原文读者对译文的理解和感受是一样的。

第二节 专业术语

下表所列词语是工程翻译人员在翻译过程中经常碰到的专业词语。

序号	中文	英文
1	100 年一遇	once in 100 years
2	安全储备	safety reserve
3	安全监测	safety monitoring
4	安全监测仪器	safety monitoring instruments
5	安全系数	safety factor
6	岸边绕渗	by-pass seepage around bank slope
7	岸边溢洪道	river-bank spillway
8	岸墙	abutment wall
9	岸塔式进水口	bank-tower intake
10	坝顶高程	crest elevation
11	坝基扬压力	uplift pressure of dam foundation
12	坝肩	dam abutment
13	坝前淤积	siltation in front of the dam
14	坝轴线	dam axis
15	薄壁堰	sharp-crested weir
16	薄拱坝	thin-arch dam
17	本构模型	constitutive model
18	鼻坎	bucket
19	比尺	scale
20	比降	gradient
21	闭门力	closing force
22	边墩	side pier
23	边界层	boundary layer
24	边墙土压力	earth pressure of side wall
25	变形监测	deformation monitoring

续表

序号	中文	英文
26	变中心角变半径拱坝	variable angle and radius arch dam
27	标准贯入试验击数	number of standard penetration test
28	标准化管理	standardized management
29	冰压力	ice pressure
30	不均匀沉降裂缝	differential settlement crack
31	不平整度	irregularity
32	财务决算和审计	financial final accounts and audit
33	侧槽溢洪道	side channel spillway
34	侧轮	side roller
35	侧收缩系数	coefficient of side contraction
36	测缝计	joint meter
37	插入式连接	insert type connection
38	差动式鼻坎	differential bucket
39	掺气	aeration
40	掺气槽	aeration slot
41	掺气减蚀	cavitation control by aeration
42	厂房顶溢流	spill over power house
43	沉积物管理概念时间表	sediment management concept schedule
44	沉降	settlement
45	沉井基础	sunk shaft foundation
46	沉沙池	sediment basin
47	沉沙建筑物	sedimentary structure
48	沉沙条渠	sedimentary channel
49	沉陷缝	settlement joint
50	沉陷观测	settlement observation
51	衬砌的边值问题	boundary value problem of lining
52	衬砌计算	lining calculation
53	衬砌自重	dead-weight of lining

续表

序号	中文	英文
54	冲沟流量（横流）	gully discharge
55	冲击波	shock wave
56	冲沙闸	flush sluice
57	冲沙隧道	flushing tunnel
58	冲刷坑	scour hole
59	冲淤	erosion and deposition
60	抽排措施	pump drainage measure
61	抽水蓄能电站厂房	pump-storage power house
62	出口段	outlet section
63	传感器	sensor
64	船闸	lock
65	垂直位移	vertical displacement
66	大头坝	massive-head dam
67	单层衬砌	monolayer lining
68	单级船闸	lift lock
69	单线船闸	single line lock
70	挡潮闸	tide sluice
71	挡水建筑物	retaining structure
72	导流隧洞/隧道	diversion tunnel
73	导墙	guide wall
74	倒虹吸管	inverted siphon
75	倒悬度	overhang degree
76	地下水位	groundwater level
77	灯泡贯流式水轮发电机组	bulb tubular turbine generator
78	等半径拱坝	constant radius arch dam
79	等中心角变半径拱坝	constant angle and variable radius arch dam
80	底流消能	energy dissipation by hydraulic jump
81	底缘	bottom edge

续表

序号	中文	英文
82	电站厂房	station workshop
83	电站厂房安装间	workshop installation room of the power station
84	电站装机	installed capacity of the power station
85	调压井	surge shaft
86	独立发电商	IPP（independent power producer）
87	发电	power generation
88	防洪	flood control
89	防洪标准	flood control standard
90	防洪库容	flood control storage
91	防洪限制水位（坝前水位）	flood control limit level
92	非汛期	non-flood season
93	分级运行水位	stage operating water level
94	分界流量	demarcation flow
95	分蓄洪区	flood diversion and storage area
96	负荷率	load rate
97	概算总投资	estimated total investment
98	钢筋	rebar
99	高程坡面	elevation slope
100	工程档案	project archive
101	工程简介	project introduction
102	观测结果	observation result
103	灌溉	irrigation
104	灌溉取水口	irrigation intake
105	国家输配电公司	National Transmission and Dispatch Company（NTDC）
106	国家重大水利工程	national major water conservancy project
107	航道	waterway
108	航运	shipping

续表

序号	中文	英文
109	河床变形	riverbed deformation
110	河道主流区	mainstream area of the river
111	后汛期	after rainy season
112	滑坡体	landslide
113	滑舌	landslide tongue
114	环境保护	environmental protection
115	环境量监测	monitoring of environmental quantity
116	混流式水轮机	francis turbine
117	混凝土重力坝	concrete gravity dam
118	基本情况	basic information
119	基荷发电	base-load generation
120	基岩变形	bedrock deformation
121	技术鉴定	technical appraisal
122	剪切裂缝	shear crack
123	江西省水利厅	Jiangxi Provincial Water Resources Department
124	降雨量	precipitation
125	铰链式闸门	flap gate
126	接缝开合度	joint degree of opening
127	径流式	run-of-river
128	竣工验收	completion acceptance
129	库容	storage capacity
130	拉张裂缝	tensile crack
131	累计发电	cumulative power generation
132	锚杆	anchor bolt
133	锚固力	anchorage force
134	锚索	anchor cable
135	泥沙管理系统	sediment management system
136	排水孔	drainage hole

续表

序号	中文	英文
137	取水许可	water intake permit
138	绕坝渗流	seepage around the dam
139	人工位移观测	manual displacement observation
140	上游水库	upstream reservoir
141	设计流量	design discharge
142	设计最大吨位	maximum designed tonnage
143	渗流监测	seepage monitoring
144	生态供水	ecological water supply
145	省重点水利工程	provincial key water conservancy project
146	数据采集设备	data acquisition equipment
147	双回输电线路	double circuit transmission line
148	水库	reservoir
149	水力学监测	hydraulics monitoring
150	水利工程标准化管理	standardized management of hydraulic engineering
151	水利工程标准化管理考核	water conservancy project standardization management assessment
152	水利枢纽工程	water conservancy project
153	水流流量	flow quantity
154	水流流速	flow velocity
155	水轮发电机组	hydropower generator set
156	水平位移	horizontal displacement
157	水土保持	soil and water conservation
158	速度	velocity
159	碎核桃壳	ground walnut shells
160	特许经营期	concession period
161	完好率	perfectness rate
162	尾水下泄	tailwater drains
163	喜马拉雅地区	Himalayan region

续表

序号	中文	英文
164	细颗粒泥沙	fine sediment
165	下泄生态流量	underflow ecological discharge
166	下泄溢洪道湾	underflow spillway bay
167	消防	fire protection
168	消力池	stilling basin
169	效益情况	benefit situation
170	泄洪洞	flood discharge tunnel
171	泄水闸	drainage lock
172	星辰水电有限公司 （Star Hydro Power Limited）	SHPL
173	兴利	good fortune
174	蓄水位	storage water level
175	悬沙荷载	suspended sediment load
176	压力水管	penstock
177	压力竖井	vertical pressure shaft
178	压应力	compressive stress
179	堰	weir
180	一度/千瓦时	kilowatt hour
181	移民安置	resettlement of immigrants
182	异常数据	abnormal data
183	溢洪道海湾溢流	overflow spillway bay
184	分水槽	bypass tunnel
185	引水隧洞	headrace tunnel
186	应急补水	emergency water supply
187	鱼道	fishway
188	过顶围堰	overtopped cofferdam
189	闸墩	pier
190	重现期	return period

续表

序号	中文	英文
191	主汛期	main flood season
192	自然沉降池	natural settling pool
193	最优混合除砂机系统（OHDS）	optimal hybrid de-sander system（OHDS）

第三节　翻译案例

案例一：中译英
原文：

×××水利枢纽工程简介

一、基本情况

×××水利枢纽工程位于赣江中游××县××镇上游峡谷河段。该工程为国家172项重大水利工程、省重点水利工程，也是迄今我省投资最大的水利枢纽工程。

××水利枢纽工程水库正常蓄水位46.0米；水库总库容××亿立方米，防洪库容××亿立方米；电站装机××兆瓦，共装9台40兆瓦的灯泡贯流式水轮发电机组；船闸设计最大吨位1000吨。

枢纽工程总布置为：河道主流区布置泄水闸，泄水闸左侧为船闸，右侧为电站厂房，左、右两岸为混凝土重力坝，坝身布置左、右岸灌溉取水口，鱼道布置在电站厂房安装间坝段右侧。坝轴线总长××米。枢纽主要建筑物沿坝轴线从左至右依次为：左岸混凝土重力坝，长××米（包括左岸灌溉总进水闸）；船闸段，长××米；门库坝段，长××米；18孔泄水闸，长××米；厂房坝段，长××米（其中安装间长××米与重力坝重合）；右岸混凝土重力坝，长××米（包括右岸灌溉总进水闸及鱼道）。设计坝顶高程××米。泄水闸墩顶高程××米，最大坝高××米。重力坝段最大坝高××米，门库坝段最大坝高××米，厂房坝段最大坝高××米。

××水利枢纽工程于×××年××月开始施工准备，×××年××月××日一期枯水围堰、枢纽主体工程正式开工。×××年××月，末台水轮发电机组具备发电条件,枢纽主体工程提前5个月完工。×××年××月至×××年××月,工程陆续通过水土保持、消防、环境保护、取水许可、安全、移民安置、工程档案等专项验收。×××年××月,工程通过竣工验收技术鉴定、竣工财务决算和审计。×××年××月××日,工程通过江西省水利厅组织的竣工验收。

工程设计概算总投资××亿元,竣工验收时实际完成投资××亿余元,目前尚有少量竣工验收尾工在建。

二、效益情况

××水利枢纽工程是一座以防洪、发电、航运为主,兼顾灌溉等综合利用的大型水利枢纽工程。工程建成后,与××分蓄洪区配合使用,可使下游的××市防洪标准由100年一遇提高到200年一遇,赣东大堤的防洪标准由50年一遇提高到100年一遇,多年平均发电量××亿度,改善上游航道××千米,并为下游沿江农田灌溉和应急补水创造条件。

截至×××年××月底,工程累计发电已超××亿度,效益十分显著。

三、安全监测情况

××水利枢纽工程大坝安全监测项目主要有:(1)水力学监测,包括鱼道水流流速及流量等;(2)环境量监测,包括大坝上下游水位、库(坝)区气温、降雨量、坝前淤积、下游冲淤及河床变形等;(3)变形监测,包括水平位移及垂直位移、倾斜度、混凝土坝接缝开合度、基岩变形等;(4)渗流监测,包括坝基扬压力、绕坝渗流、地下水位监测等;(5)应力、应变及温度监测,包括混凝土应力和应变、钢筋应力、锚杆应力、锚索锚固力、坝体混凝土及坝基温度、船闸边墙土压力等。

××水利枢纽工程大坝共安装969支安全监测仪器。截至×××年××月,接入自动化监测系统的传感器失效的共计31支,完好率为97%。自动化系统设备完好率为100%。枢纽大坝安全监测工作实行水利工程标准化管理,观测方法及频次满足规程规范要求,遇洪水、大雨、台风、水库水位骤变、水库蓄水等特殊情况加密测次,定期进行观测资料的整理分析并及时对异常数据进行分析和排查,指导研判大坝安全状况。同时保证数据采集设备的正常运行,确保监测数据及时上传至工作站,数据上传延迟时间不超过24小时。

目前,枢纽各方面监测情况显示,枢纽大坝、电站、船闸等建筑物运行状态正常。枢纽工程的工作状况和安全性评价如下：

（1）左右岸挡水坝及门库坝基扬压力、基岩变形、温度、接缝变形及混凝土应变观测结果正常。

（2）船闸坝基扬压力、接缝变形、钢筋应力、混凝土应变、压应力测值正常，锚索应力变化比较平稳，船闸观测结果正常。

（3）泄水闸坝基扬压力、基岩变形、接缝变形、闸基温度、应力应变、钢筋应力、锚杆应力、闸墩倾斜观测结果正常。

（4）厂房坝基扬压力、基岩变形、接缝变形、混凝土温度、应力应变、钢筋应力观测结果正常。

（5）大坝表面水平位移、表面垂直位移观测结果正常。

（6）大坝绕坝渗流压力观测结果正常。

（7）左右岸边坡地下水位、右岸边坡锚索应力、左岸边坡内部位移观测结果正常。

××××年××月至××月初，坝址区经历了三轮持续强降雨。右岸坝肩原滑坡体覆盖层因渗水饱和，导致重新活动，原滑坡体后缘拉张裂缝开度增大，下游北翼出现连续剪切裂缝，滑舌前缘已向下发展至51.2—61.0米高程坡面。通过采取用黏土封填右岸滑坡体拉张裂缝、重新钻设深层排水孔等处理措施后，连续人工位移观测显示，右岸坝肩滑坡体裂缝已趋于稳定。

译文：

Introduction of ×× Water Conservancy Project

1. Basic Information

×× Water Conservancy Project is located in the upper gorge reach of ×× Town, ×× County, which is in the middle reaches of Ganjiang River. It is one of the 172 national major water conservancy projects and provincial key water conservancy projects, as well as the most invested water conservancy project in Jiangxi Province so far.

The normal storage water level of ×× Water Conservancy Project reservoir is 46.0 meters; The total storage capacity of the reservoir is ×× billion cubic meters, and the flood control storage is ×× million cubic meters. The installed capacity of the power station is × × MW, including 9 sets of bulb tubular turbine generator with 40 MW each. The maximum designed tonnage of the lock is 1000 tons.

The general arrangement of the conservancy project is as follows: the drainage lock is in the mainstream area of the river, with the ship lock is on the left side and the station workshop is on the right side. The concrete gravity dams are on both banks of the river, on which arranged the irrigation intakes, and the fishway is arranged on the right side of the dam section of workshop installation room of the power station. The total length of the dam axis is × × m, and the main buildings of the project from left to right along the dam axis are: the concrete gravity dam on the left bank (× × m long, including the total irrigation inlet sluice of left bank), ship lock section (× × m long), gate chamber dam section(× × m long), 18-hole drainage lock (× × m long), workshop dam section (× × m long, among it the installation room is × × m long, coinciding with the gravity dam), and the concrete gravity dam on the right bank (× × m long, including the main irrigation inlet sluice of right bank and fishway); Designed crest elevation is of × × m, the elevation of the drainage gate pier is × × m, the maximum dam height is × × m, the maximum height of gravity dam section is × × m, the maximum height of gate chamber dam section is × × m, and the maximum height of workshop dam section is × × m.

The construction preparation of × × Water Conservancy Project began in date × ×. On date × ×, the first phase of the starter cofferdam and the main body of the project were officially started. In date × ×, the last hydropower generator set was ready for power generation, and the main body of the project was completed 5 months in advance. From date × × to date × ×, the project passed special acceptance checks, such as soil and water conservation, fire protection, environmental protection, water intake permit, safety, resettlement of immigrants, and project archives. In date × ×, the project passed the technical appraisal, financial final accounts and audit of completion acceptance. On date × ×, the project passed the completion acceptance organized by Jiangxi Provincial Water Resources Department.

The estimated total investment of the project is × × billion yuan, but the actual investment is over × × billion yuan at the time of completion acceptance. At present, a small amount of completion acceptance work is still under construction.

2. Benefit Situation

× × Water Conservancy Project is a multipurpose large-scale project which focuses on flood control, power generation, shipping, and considers irrigation as well. After the completion of the project, it is used cooperatively with × × flood diversion and storage area, so that the flood control standard of downstream × × City can be raised from once in 100 years to once in 200 years, and the standard of Gandong levee can be raised from once in 50 years to once in 100 years. The project obtains an annual average power generation of × × billion kWh, improves × × km of upstream waterway and creates conditions for irrigation and emergency water supply to the farmland downstream banks along the river.

By the end of date × ×, the cumulative power generation by the project had exceeded × × billion kWh, with remarkable benefits.

3. Safety Monitoring Situation

The dam safety monitoring programs of × × Water Conservancy Project mainly include: (1) Hydraulics monitoring, including the flow velocity and quantity of the fishway; (2) Monitoring of environmental quantity, including the water level of up and down stream of the dam, temperature in the reservoir (dam) area, precipitation, siltation in front of the dam, downstream erosion and deposition, riverbed deformation, etc.; (3) Deformation monitoring, including horizontal displacement and vertical displacement, incline, concrete dam joint degree of opening, bedrock deformation, etc.; (4) Seepage monitoring, including uplift pressure of dam foundation, seepage around the dam, groundwater level monitoring, etc.; (5) Stress, strain and temperature monitoring, including stress and strain of concrete, stress of rebar, stress of anchor bolt, anchorage force of anchor cable, temperature of concrete in dam body and foundation, earth pressure of side wall of ship lock, etc. .

A total of 969 safety monitoring instruments have been installed in the dam of × × Water Conservancy Project. As of date × ×, 31 sensors connected to the automatic monitoring system had failed with a serviceability rate of 97%, and the automatic system equipment perfectness rate is 100%. The safety monitoring work of the dam implements the standardized management of hydraulic engineering, the observation methods and frequency meet the requirements of the regulations and specifications, and the observation frequency shall be increased in the case of flood, heavy rain, typhoon, sudden change of reservoir water level, reservoir filling and other special circumstances. The observation data is sorted out and analyzed regularly, and abnormal data shall be analyzed and checked in time to guide the judgment of the dam safety status. At the same time, the monitoring system should ensure the normal operation of data acquisition equipment and ensure that the monitoring data is timely uploaded to the workstation which is not allowed delay more than 24 hours.

At present, according to the monitoring situation of all aspects of the project, the dam, power station, ship lock and other buildings are running normally. The working status and safety evaluation of the project are as follows:

(1) The observation results of foundation uplift pressure, bedrock deformation, temperature, joint deformation and concrete strain of retaining dams on both banks and gate chamber dam are normal.

(2) The measure values of foundation uplift pressure, joint deformation, reinforcement stress, concrete strain and compressive stress of lock dam are normal, the stress changes of anchor cable are relatively stable, and the observation results of the lock are normal.

(3) The observation results of foundation uplift pressure, bedrock deformation, joint deformation, temperature of gate foundation, stress-strain and reinforcement stress, stress of anchor bolt and tilt of pier of sluice dam are normal.

(4) The observation results of foundation uplift pressure, bedrock deformation, joint deformation, concrete temperature, stress-strain and reinforcement stress of workshop dam are normal.

(5) The observation results of horizontal and vertical displacement of the dam surface are normal.

(6) The observation result of seepage pressure around the dam is normal.

(7) The observation results of groundwater level of left and right bank slope, anchor cable stress of right bank slope and internal displacement of left bank slope

are normal.

From date × × to early date × ×, the dam site experienced three rounds of continuous heavy rainfall. The overburden of the original landslide on the right bank dam abutment reactivated because of saturation with water seepage, which led to an increase of tensile crack at the rear edge of the original landslide, and the continuous shear crack appeared on the downstream north wing. The leading edge of the landslide tongue had developed downward to an elevation slope of 51.2 – 61.0 m. After clay sealing the tensile crack, re-drilling deep drainage holes and other treatment measures to the right bank landslide, the continuous manual displacement observation shows that the slip mass cracks of the right bank dam abutment have tended to be stable.

案例二:英译中
原文:

Patrind Hydropower Project

The run-of-river Patrind Hydropower Project has been constructed on river Kunhar. With a capacity of 147 MW (Net), the Project shall generate, on average, 632 GWh of electricity annually during the concession period of 30 years. Since the project has been developed by SHPL as an IPP, the SHPL has entered into a 30-year Power Purchase Agreement with National Transmission and Dispatch Company (NTDC), Pakistan's grid system operator, for the sale of electricity generated from the project.

The project is located on the boundary of Azad Jammu & Kashmir (AJ&K) and District Abbottabad of Pakistan, near the city of Muzaffarabad The majority of the Project structures, including the powerhouse, are located in the territory of AJ&K. However, the diversion tunnel, flushing tunnel and a part of the weir are located within territorial limits of District Abbottabad.

The weir side of the project can be reached through Boi Road on right side of river Kunhar at a distance of approximately 12.3 km from Garhi Habibullah, a small town in District Mansehra. The powerhouse side of the project is accessible from lower Chattar, Muzaffarabad where a class 70 bridge has been constructed across river Jhelum as a part of the project for access to the powerhouse.

The project diverts the waters of river Kunhar through a weir, located near village Patrind, and a left bank conveyance system of headrace tunnel to the right bank of river Jhelum, near the city of Muzaffarabad, where a powerhouse has been built. The natural difference of elevation between river Kunhar and river Jhelum, along with a 43.5 m high weir provides a suitable head to set up a 147 MW power project.

Major project components include a weir structure, an upstream concrete cofferdam and a flushing tunnel for sediments flushing, an intake structure slightly upstream of the weir leading to the headrace tunnel. The headrace tunnel passes through a ridge towards river Jhelum. It connects to an underground surge shaft via link tunnel and further opens into vertical pressure shaft, which connects with a horizontal pressure tunnel. The penstock divides into three manifolds conveying design discharge to a surface type powerhouse located on the right bank of the river Jhelum.

Since river Kunhar is a river with high concentration of sediments, a proper and effective sediment management scheme has been the priority for SHPL. An Optimal Hybrid Desander System (OHDS)/Rearranged Sandtrap has been preferred for the project over any scheme involving surface or underground sandtrap. OHDS is composed of a flushing tunnel and a modified pool. Flushing tunnel of suitable dimensions (tested in numerical and physical modelling) is provided at upstream of the concrete cofferdam, located upstream of the weir, to flush out the sediments from the reservoir when the inflow is more than 200 m^3/s. The pond between upstream cofferdam and the weir will be used as the natural sandtrap for settling down particles. The settled particles shall be flushed out using under-sluice radial gates provided in the weir structure. The OHDS scheme has been tested through numerical modelling by HR Wallingford and physical modelling at ETH, Zurich, Switzerland.

Sediment management system also relies on proper reservoir operation procedure. SHPL has prepared a comprehensive plan for reservoir operation, which has also been made part of the operating procedure of the project in consultation with the power purchaser.

While the OHDS has been shown to be effective in managing sediments at the reservoir, SHPL has also opted for HVOF coating for its three 50 MW francis turbines as added protection against adverse impacts of any particles flowing into the power intake.

The project is connected to Pakistan's national grid through 132 kV double circuit transmission line.

译文：

<h2 style="text-align:center">帕特灵水电站项目</h2>

帕特灵水电站是修建在库哈尔河上的一座径流式水电站。项目特许期为30年，年平均发电量为632吉瓦时，装机容量为147兆瓦（净）。该项目是由星辰水电有限公司（Star Hydro Power Limited）作为独立发电商（independent power producer）开发的，SHPL已与巴基斯坦电网系统运营商——国家输配电公司（NTDC）签订了一份为期30年的电力购买协议，以出售该项目产生的电力。

该项目位于巴基斯坦的阿扎德查谟和克什米尔（AJ&K）及阿伯塔巴德地区的边界，靠近穆扎法拉巴德市。包括发电站在内的大部分项目结构位于AJ&K境内。而导流隧道、冲砂隧道和部分堰位于阿伯塔巴德地区的地域范围内。

工程的堰侧可通过库哈尔河右侧的博伊路到达，距离曼塞赫拉区的加希哈比布拉小镇约12.3千米。发电厂房一侧可以从穆扎法拉巴德市的卡塔下游进入，该地一座70级的桥梁横跨杰勒姆河，作为水电站的一部分用于进入厂房。

该项目通过帕特灵村附近的堰和引水隧洞左岸输送系统，将库哈尔河的水引流到穆扎法拉巴德市附近的杰勒姆河右岸，并在那里建造了一座发电站。库哈尔河和杰勒姆河之间的自然高程差，以及43.5米高的堰，为建设一座147兆瓦的电力工程提供了合适的条件。

本项目主要工程包括一个堰结构，一个上游混凝土围堰，一个用于冲刷沉积物的冲砂隧道，一个与引水隧洞相连、略高于堰的取水结构。引水隧洞穿过山脊通向杰勒姆河，通过连接隧道与地下调压井相通，进一步贯通压力竖井，并与水平压力隧道相连。压力管分为三组管汇，将设计流量输送到位于杰勒姆河右岸的地面厂房。

由于库哈尔河是一条含沙量高的河流，出具一套合理有效的治沙方案一直是星辰水电有限公司的首要任务。在该项目中,最优混合除砂系统(OHDS)/重排式捕砂器优于任何地面或地下捕砂器的方案。最优混合除砂系统由冲洗通道和修正池组成。混凝土围坝的上方(堰上方)设有适当尺寸的冲刷隧道(进行了数值和物理模拟试验),在水流量超过 200 米³/秒时可将水库中的沉积物冲刷出去。上游围坝与堰之间的池塘将用作天然沙池,沉淀颗粒。沉淀的颗粒将通过堰结构内的底部弧形闸门排出。最优混合除砂系统方案已通过英国水力研究院的数值模拟和瑞士苏黎世联邦理工学院的物理模拟测试。

泥沙管理系统还依赖于合理的水库运行程序。星辰水电有限公司已经制订了一个全面的水库运行方案,与电力采购方协商决定将其作为本项目运行程序的一部分。

虽然已经证明最优混合除砂系统在管理水库沉积物方面是有效的,但星辰水电有限公司还为三台 50 兆瓦的混流式水轮机提供了额外的超音速火焰涂层防护,以防止任何颗粒进入电站进水口带来的不利影响。

该项目已通过 132 千伏双回输电线路与巴基斯坦国家电网相连。

第四节　翻译评析

一、词法层面

(一)数字、单位、时间、地名、组织单位等的翻译

工程文本中通常包含大量的数字、单位、时间、编号等信息,在翻译时应忠实转录,绝不能出现错误。金额和数字的表达,必须符合目标语的语言习惯;数字单位翻译为目标语中的相应表达,不可随意变换;地名、组织单位等需要进行认真查证后再进行翻译。

【例1】原文:××水利枢纽工程水库正常蓄水位××米;水库总库容××亿立方米,防洪库容××亿立方米。

译文:The normal storage water level of × × Water Conservancy Project reservoir is × × meters; The total storage capacity of the reservoir is × × billion cubic meters, and the flood control storage is × × million cubic meters.

分析:例 1 在翻译时为了保持数字的精确,原文小数点后面的 0 也依然保

留,同时不改变原文的水位和库容单位。但由于中英文数字表达方式的差异,在翻译时应注意确认。

【例2】原文:The project is located on the boundary of <u>Azad Jammu & Kashmir (AJ&K)</u> and <u>District Abbottabad of Pakistan</u>, near the city of <u>Muzaffarabad</u>.

译文:该项目位于巴基斯坦的<u>阿扎德查谟和克什米尔(AJ&K)</u>及<u>阿伯塔巴德地区</u>的边界,靠近<u>穆扎法拉巴德市</u>。

分析:英语中不常见的地名、人名、组织名称同样需要译者仔细查询以确认其中文译名。若没有统一的既定译法,可选择意译法或音译法,或者"意译 + 音译"结合的方法。

(二)术语的翻译

工程翻译涉及大量的专业术语和非专业词语。通常来说,专业术语的词义为一一对应,在进行译前准备工作时,译者应根据所翻译文本的类型准备相应的英汉专业术语库或术语表辅助翻译。而对于术语库或者词典中找不到恰当词义的词语,译者需要根据上下文内容和逻辑关系,从该词的根本含义出发,结合相关专业去判断词义。因此,译者需对所翻译的专业术语和基础知识进行一定的了解和熟悉,才能使译文逻辑正确、用词专业。

【例3】原文:<u>电站装机</u>××兆瓦,共装 9 台 40 兆瓦的<u>灯泡贯流式水轮发电机组</u>。

译文:The <u>installed capacity of the power station</u> is × × MW, including 9 sets of <u>bulb tubular turbine generator</u> with 40 MW each.

分析:电站装机(installed capacity of the power station),全称"发电厂装机容量",指水电站中所装有的全部汽轮或水力发电机组额定功率的总和,是表征一座水电站建设规模和电力生产能力的主要指标之一。

"灯泡贯流式水轮发电机"译为"bulb tubular turbine generator"。在贯流式水轮发电机中,有一种将发电机组安装在外形酷似灯泡的密封壳体中,将水轮机安装于灯泡插口处的机组,被称为灯泡贯流式机组。其适用的水头范围较广,工作效率高,因而备受青睐。

【例4】原文:The headrace tunnel passes through a ridge towards river Jhelum. It connects to an underground <u>surge shaft</u> via link tunnel and further opens into <u>vertical pressure shaft</u>, which connects with a horizontal pressure tunnel.

译文:引水隧洞穿过山脊通向杰勒姆河,通过连接隧道与地下<u>调压井</u>相通,进一步贯通<u>压力竖井</u>,并与水平压力隧道相连。

分析:"surge shaft"译为调压井。调压井也称压力井,是用于水电站等地起调节水压作用的机械设备。如果引水式水电站的引水管道的总长度较长,就常常要在隧洞与压力管道交界处设置调压室。调压室可以是从山体中开挖出来的井式结构(称为调压井),可以是高出地面的一个塔(称为调压塔),也可以是某种混合结构。发电站的引水管道较长,当机组运行中突然甩负荷关闭导叶时,由于水流的惯性作用,会有很大的水锤效应,易损毁发电设备。如无调压井,水锤会击毁导水叶和其他过流部件。调压井的作用就是让水锤有一个释放的通道,以减小过流部件的压力。

"vertical pressure shaft"译为压力竖井,其功能与调压井相同。

(三)缩略词的翻译

【例5】原文:Since the project has been developed by SHPL as an IPP, the SHPL has entered into a 30-year Power Purchase Agreement with National Transmission and Dispatch Company (NTDC), Pakistan's grid system operator, for the sale of electricity generated from the project.

译文:该项目是由星辰水电有限公司(Star Hydro Power Limited)作为独立发电商(independent power producer)开发的,SHPL已与巴基斯坦电网系统运营商——国家输配电公司(NTDC)签订了一份为期30年的电力购买协议,以出售该项目产生的电力。

分析:英译汉的过程中,英语缩略词也是常见的翻译难点。遇到不常见的缩略词,且上文未说明缩略词的含义时,译者应认真查询缩略词的全称及其内涵,并选择合适的词语进行翻译。本句中的缩略词"SHPL"和"IPP",译者经过多方查询得知其全称分别为Star Hydro Power Limited和independent power producer,从而准确翻译出其含义。

(四)插入语的翻译

【例6】原文:SHPL has entered into a 30-year Power Purchase Agreement with National Transmission and Dispatch Company (NTDC), <u>Pakistan's grid system operator</u>...

译文:SHPL已与巴基斯坦电网系统运营商——国家输配电公司(NTDC)签

订了一份为期30年的电力购买协议……

分析：英语中常用插入语，而直译为中文插入语会影响阅读的连贯性。为不影响主句的阅读体验，译者在翻译插入语时可以采用两种方式：1.改变语序或增加从句的方式将插入语整合到主句中；2.采用破折号或者括号作为插入部分。

（五）词义选择

中英文的词语含义并非完全对等。相比而言，英语的词义比较灵活，词义范围也相对丰富多变，词义对上下文的依赖性较大；汉语的词义则比较严谨，词义范围相对较窄且比较精确固定，词义对上下文的依赖性较小。因此，工程翻译中的词义应结合工程英语专业知识来确定。

【例7】原文：SHPL has entered into a 30-year Power Purchase Agreement with National Transmission and Dispatch Company (NTDC) ...

译文：SHPL已与巴基斯坦电网系统运营商——国家输配电公司（NTDC）签订了一份为期30年的电力购买协议……

分析："enter into"直译为"进入"。根据上下文可知，本句中"enter into"应指签订合作协议，因此采用意译的方法。

【例8】原文：The project diverts the waters of river Kunhar through a weir, located near village Patrind to the right bank of river Jhelum, near the city of Muzaffarabad.

译文：该项目通过帕特灵村附近的堰将库哈尔河的水引流到穆扎法拉巴德市附近的杰勒姆河右岸。

分析：本句中"divert"可与汉语中的"转移"对应，但是从水利工程专业以及上下文的角度来看，翻译为"引流"显然更专业和贴切。

【例9】原文：The natural difference of elevation between river Kunhar and river Jhelum, along with a 43.5m high weir provides a suitable head to set up a 147 MW power project.

译文：库哈尔河和杰勒姆河之间的自然高程差，以及43.5米高的堰，为建设一座147兆瓦的电力工程提供了合适的条件。

分析：本句中"head"很容易被理解为"源头"。然而，根据前文，较大的高度落差应为建设水电站的"条件"而非"源头"。若将其翻译为"为建设电力工程提供了……的源头"显然用词不够准确。

在案例二中，第一段在简单介绍之后，后文均用"project"指代帕特灵水电站项目，翻译时可适当变化词语，采用"工程、项目、水电站"等词语翻译"project"，避免多次重复。

（六）词性转换

【例10】原文：电站装机××兆瓦，共装9台40兆瓦的灯泡贯流式水轮发电机组。

译文：The installed capacity of the power station is ×× MW, including 9 sets of bulb tubular turbine generator with 40 MW each.

分析：本处使用了词性转换的翻译技巧，采用动名词"including"做状语翻译原文的"副词 + 动词"的偏正结构，起到简化句子结构、增强句子间逻辑关系的作用，主要用于翻译从属的动作。

【例11】原文：工程设计概算总投资××亿元，竣工验收时实际完成投资××亿余元，目前尚有少量竣工验收尾工在建。

译文：The estimated total investment of the project is ×× billion yuan, but the actual investment is over ×× billion yuan at the time of completion acceptance. At present, a small amount of completion acceptance work is still under construction.

分析：工程英语多采用名词化结构使行文严谨，中译英时也应符合英语的语言特点。本句采用"介词 + 名词"结构"under construction"翻译原文中的动词"在建"，这是中译英时动词词性转换的常见方法。

（七）多种词汇翻译技巧整合

【例12】原文：Major project components include a weir structure, an upstream concrete cofferdam and a flushing tunnel for sediments flushing, an intake structure slightly upstream of the weir leading to the headrace tunnel. The headrace tunnel passes through a ridge towards river Jhelum. It connects to an underground surge shaft via link tunnel and further opens into vertical pressure shaft, which connects with a horizontal pressure tunnel.

译文：本项目主要工程包括一个堰结构，一个上游混凝土围堰，一个用于冲刷沉积物的冲砂隧道，一个与引水隧洞相连、略高于堰的取水结构。引水隧洞穿过山脊通向杰勒姆河，通过连接隧道与地下调压井相通，进一步贯通压力竖井,并与水平压力隧道相连。

分析:"major"的真正修饰对象是"component"。中文中修饰语与被修饰对象应尽量靠近,译文的形容词"主要的"应移到"工程"之前。

"upstream"一般译为"上游的/地",但由于前面还有一个副词"slightly",连续翻译前置状语较为冗长,可根据其表达的内涵直接翻译为"略高于"。

定语"leading to"采用转换词性的方式翻译为介词短语"与……相连",而非"通向……",使译文更为流畅、准确。

本句具有多处与"连接"相关的表达——"connect""link tunnel"和"open into",翻译时应选择多样化的词语避免重复。例 12 中就采用具体形象的动词"open into"来表达"连接"这个抽象概念,翻译时可进行引申,将其译为"贯通"。

二、句法层面

(一)长句的处理

【例 13】原文:枢纽工程总布置为:河道主流区布置①泄水闸,泄水闸②左侧为船闸、右侧为电站厂房,左、右两岸为混凝土重力坝,坝身布置左、右岸③灌溉取水口,鱼道布置在电站厂房安装间坝段右侧。坝轴线总长××米,枢纽主要建筑物沿坝轴线从左至右依次为:左岸混凝土重力坝长××米④(包括左岸灌溉总进水闸),船闸段长××米,门库坝段长××米,18 孔泄水闸长××米,厂房坝段长××米(其中安装间长××米与重力坝重合),右岸混凝土重力坝长××米(包括右岸灌溉总进水闸及鱼道);设计坝顶高程××米。泄水闸墩顶高程××米,最大坝高××米,重力坝段最大坝高××米,门库坝段最大坝高××米,厂房坝段最大坝高××米。

译文:The general arrangement of the conservancy project is as follows: the drainage lock is in the mainstream area of the river, with the ship lock is on the left side and the station workshop is on the right side. The concrete gravity dams are on both banks of the river, on which arranged the irrigation intakes, and the fishway is arranged on the right side of the dam section of workshop installation room of the power station. The total length of the dam axis is × × m, and the main buildings of the project from left to right along the dam axis are: the concrete gravity dam on the left bank (× × m long, including the total irrigation inlet sluice of left bank), ship lock section (× × m long), gate chamber dam section(× × m long), 18-hole drainage lock (× × m long), workshop dam section (× × m long, among it the installation room is × × m long, coinciding with the gravity dam), and the concrete gravity dam on the right bank (× × m long, including the main irrigation inlet sluice of right bank and fishway); Designed crest elevation is of × × m, the elevation of the drainage gate pier is × × m, the maximum dam height is × × m, the maximum height of gravity dam section is × × m, the maximum height of gate chamber dam section is × × m, and the maximum height of workshop dam section is × × m.

分析:本段落有多处重复出现同义词语,翻译时可采用省略、替代、转换句子结构等方式,实现语篇衔接流畅,避免重复的目的。

①前文已将"布置"翻译为"arrangement",此处仅以介词"in"表示河道主流区和泄水闸的位置关系,避免重复。

②工程翻译以简洁直观为主,本句省略重复提及的主语"泄水闸",以 with 结构表示船闸和电站厂房与泄水闸的关系。

③为避免句子过于分散,本句采用"on which"表示混凝土重力坝和灌溉取水口的位置关系,并省略重复提及的"左右两岸"。

④本部分为工程各部分的高度值,翻译时采用括号内补充说明的方式将"……长"结构简略翻译,避免句型重复累赘。

【例14】原文:××水利枢纽工程……多年平均发电量××亿度,改善上游航道××千米,并为下游两岸沿江农田灌溉和应急补水创造条件。

译文:× × Water Conservancy Project... The project obtains an annual average power generation of × × billion kWh, improves × × km of upstream waterway and creates conditions for irrigation and emergency water supply to the farmland downstream banks along the river.

分析:本句最后一部分内容"多年平均发电量""改善上游航道"和"为……创造条件"均省略了同一主语"××水利枢纽工程"。在翻译时应注意补充主语,并增加动词"obtain",将"多年平均发电量××亿度"翻译为动宾结构,形成三个主语相同的并列分句,保证译文意思完整。

【例15】原文:同时保证数据采集设备正常运行,确保监测数据及时上传至工作站,数据上传延迟时间不超过 24 小时。

译文:At the same time, <u>the monitoring system</u> should ensure the normal operation of data acquisition equipment and ensure that the monitoring data is timely uploaded to the workstation which is not allowed delay more than 24 hours.

分析:原文是典型的主语省略句型,根据上下文可知,其主语为"安全检测系统"。在翻译时,应根据英语语法规则增加主语"the monitoring system"。

(二)语态的翻译

如前文所述,汉语多用主动语态,英语多用被动语态,而汉语也有多种表达被动含义的形式。英译中时应以忠实准确、通顺规范为标准选择主动或被动语态进行翻译。

【例 16】原文:The run-of-river Patrind Hydropower Project has been constructed on river Kunhar.

译文:帕特灵水电站是修建在库哈尔河上的一座径流式水电站。

分析:本句采用被动语态介绍水电站所处位置和水电站类型,翻译时转换为主动语态,添加原句缺少的宾语,符合中文阅读习惯。原句主语"Patrind Hydropower Project"有点儿长,容易偏移重点,因此翻译时将定语"run-of-river"移到宾语之前,起到平衡语句且不改变原文含义的作用。

【例 17】原文:××水利枢纽工程大坝共安装 969 支安全监测仪器。截至×××年××月,接入自动化监测系统的传感器失效的共计 31 支,完好率为 97%,自动化系统设备完好率为 100%。

译文:A total of 969 safety monitoring instruments have been installed in the dam of × × Water Conservancy Project. As of date × ×, 31 sensors connected to the automatic monitoring system had failed with a serviceability rate of 97%, and the automatic system equipment perfectness rate is 100%.

分析:本句全部采用主动变被动的翻译方法。由于主语不同,翻译时无法将所有内容整合到同一句子中,因此需要对句子进行拆分,并使用"with"结构整合分散信息。

(三)句子结构调整

【例 18】原文:通过采取用黏土封填右岸滑坡体拉张裂缝、重新钻设深层排水孔等处理措施后,连续人工位移观测显示,右岸坝肩滑坡体裂缝已趋于稳定。

译文:After clay sealing the tensile crack, re-drilling deep drainage holes and other treatment measures to the right bank landslide, the continuous manual displacement observation shows that the slip mass cracks of the right bank dam abutment have tended to be stable.

分析:原文的前置状语较长,只有"右岸坝肩滑坡体裂缝已趋于稳定"为主句。翻译时为避免状语过长影响阅读体验,将本句的主句调整为"the continuous manual displacement observation shows that..." ,实现语篇衔接和语义连贯。

【例19】原文:The majority of the project structures, including the powerhouse, are located in the territory of AJ&K.

译文:包括发电站在内的大部分项目结构位于 AJ&K 境内。

分析:英语中状语的位置非常多变,而中文只有句首状语和句尾状语,句首状语用于修饰整个句子,句尾状语通常作为一种附加说明或额外解释。翻译时,译者应根据状语的功能和中文的语法习惯灵活调整状语的位置。本句翻译时将状语移到句首,修饰主语。

【例20】原文:The project diverts the waters of river Kunhar through a weir, located near village Patrind, and a left bank conveyance system of headrace tunnel to the right bank of river Jhelum, near the city of Muzaffarabad, where a powerhouse has been built.

译文:该项目通过帕特灵村附近的堰和引水隧洞左岸输送系统,将库哈尔河的水引流到穆扎法拉巴德市附近的杰勒姆河右岸,并在那里建造了一座发电站。

分析:本句中"through"引导的状语较长,主句内容为"the project diverts the waters of river Kunhar to river Jhelum"。而中文习惯根据事情发生的时间顺序、因果逻辑等来组织串联句子,因此翻译时将状语"through"提前,将谓语和宾语置后,添加动词"将"补充句子完整性。此外,本句中插入了两个地点状语和由"where"引导的状语从句,翻译时注意辨别主语,并将其放置在恰当的位置,同时要添加连词和地点状语"并在那里"来拆分翻译,避免译文过于累赘。

(四)复杂句的翻译

【例21】原文:同时保证数据采集设备正常运行,确保监测数据及时上传至工作站,数据上传延迟时间不超过 24 小时。

译文：At the same time, the monitoring system should ensure the normal operation of data acquisition equipment and ensure that the monitoring data is timely uploaded to the workstation which is not allowed delay more than 24 hours.

分析：本句主语为"监测数据"，与主句不同，且属于次要内容，采用定语从句翻译，保持主句的主语一致，且不影响原句内容的主次关系。定语从句中用"which"代替主语，添加谓语"is not allowed"进行语态转换，从而省略原句的重复语义"上传时间"。

【例22】原文：The powerhouse side of the project is accessible from lower Chattar, Muzaffarabad where a class 70 bridge has been constructed across river Jhelum as a part of the project for access to the powerhouse.

译文：发电厂房一侧可以从穆扎法拉巴德市的卡塔下游进入，该地一座70级的桥梁横跨杰赫勒姆河，作为水电站的一部分用于进入厂房。

分析：翻译定语从句时，一般需要增加相应的名词或代词作为补充主语，注意用词的准确和逻辑的通顺。

【例23】原文：×××年××月至××月初，坝址区经历了三轮持续强降雨。右岸坝肩原滑坡体覆盖层因渗水饱和，导致重新活动，原滑坡体后缘拉张裂缝开度增大，下游北翼出现连续剪切裂缝，滑舌前缘已向下发展至51.2—61.0米高程坡面。

译文：From date × × to early date × ×, the dam site experienced three rounds of continuous heavy rainfall. The overburden of the original landslide on the right bank dam abutment reactivated because of saturation with water seepage, which led to an increase of tensile crack at the rear edge of the original landslide, and the continuous shear crack appeared on the downstream north wing. The leading edge of the landslide tongue had developed downward to an elevation slope of 51.2 – 61.0 m.

分析：对于长句应先进行句子结构分析，并查询相关专业术语的内涵，从而充分理解原文，选择恰当的翻译技巧。经过分析可知，右岸坝肩处曾经遭到滑坡破坏，后覆盖修复。三轮强降雨导致该处重新移动，引起后缘处横向拉张裂缝再度增大，下游出现连续纵向剪切裂缝，且滑坡前缘已向下滑落，形成较高坡面。原文句子结构并不复杂，但是专业术语较多，表述较为抽象，缺乏水利工程专业知识的译者理解起来较为困难。前面三个分句之间存在连续因果关系，可采用非限制性定语从句翻译，整合为一个完整句。最后一个分句主语变成"滑舌前缘"，因此作为一个单独的句子进行翻译。

【例 24】原文：With a capacity of 147 MW（Net），the project shall generate，on average，632 GWh of electricity annually during the concession period of 30 years.

译文：项目特许期为 30 年，年平均发电量为 632 吉瓦时，装机容量为 147 兆瓦（净）。

分析：本句结构较为松散，有伴随状语、插入语和时间状语，翻译时采用同主语连续并列分句，改变原句语序和句子成分，整合各部分内容，客观说明水电站的基本数据信息。

【例 25】原文：While the OHDS has been shown to be effective in managing sediments at the reservoir，SHPL has also opted for HVOF coating for its three 50 MW francis turbines as added protection against adverse impacts of any particles flowing into the power intake.

译文：虽然已经证明最优混合除砂系统在管理水库沉积物方面是有效的，但星辰水电有限公司还为三台 50 兆瓦的混流式水轮机提供了额外的超音速火焰涂层防护，以防止任何颗粒进入电站进水口带来的不利影响。

分析：长句在工程英语中使用频率很高，因为可以表达多个紧密相关的概念。长句的翻译主要存在英汉语序差异和英汉表达方式差异的问题，因此必须分析清楚每个长句的深层结构后再进行翻译。本句由句首的让步状语、主句和句尾的目的状语构成。主句结构为"SHPL has also opted for...francis turbines"，译为汉语时无法整合在同一句子中，尤其是在两个状语连续的情况下（"as added protection against..."），因此必须根据中文语法将其整合到其他分句中去。此处将"as added protection"与主句整合翻译为"额外的……防护"，将目的状语中的"against adverse impacts"进行拆分并变换语序，译为"以防止……带来的不利影响"。

【例 26】原文：Since river Kunhar is a river with high concentration of sediments，a proper and effective sediment management scheme has been the priority for SHPL. An Optimal Hybrid Desander System（OHDS）/Rearranged Sandtrap has been preferred for the project over any scheme involving surface or underground sandtrap. OHDS is composed of a flushing tunnel and a modified pool. Flushing tunnel of suitable dimensions (tested in numerical and physical modelling) is provided at upstream of the concrete cofferdam, located upstream of the weir, to flush out the sediments from the reservoir when the inflow is more than 200 m³/s.

译文：由于库哈尔河是一条含沙量高的河流，出具一套合理有效的治沙方案一直是星辰水电有限公司的首要任务。在该项目中，最优混合除砂系统（OHDS）/重排式捕砂器优于任何地面或地下捕砂器的方案。最优混合除砂系统由冲洗通道和修正池组成。混凝土围坝的上方（堰上方）设有尺寸适当的冲刷隧道（进行了数值和物理模拟试验），在水流量超过 200 米³/秒时可将水库中的沉积物冲刷出去。

分析：通常来说，单一的翻译技巧无法呈现出质量较高的译文，各种翻译技巧应在翻译中组合运用。例如，本句就采用了词义选择（provide——设有，sediment——泥沙、沉积物）、增译法（当……时）、句型转换（"to flush out..."，可将……冲刷出去）等翻译技巧。

三、语篇层面

构成文本的句子不仅要在结构上有联系，而且要在语义上有联系。因此，句子是篇章的基础，句子之间的衔接和连贯影响着整个篇章。众所周知，汉语中句与句之间的关系基本是平行的，按时间顺序或逻辑顺序排列和组合，词的连接通过语义、词序、逻辑关系和虚词来完成。在源文本为工程类文本的前提下，这种表达方式更为直接。而英语的词语连接是通过连词、各种短语和语言的形态变化来实现的，强调句式，句法结构严谨。在项目管理文本翻译中，译者应了解句子和段落之间的语义关系，以保证语篇的衔接和连贯。

【例27】原文：While the OHDS has been shown to be effective in managing sediments at the reservoir, SHPL has also opted for HVOF coating for its three 50 MW francis turbines as added protection against adverse impacts of any particles flowing into the power intake.

译文：虽然已经证明最优混合除砂系统在管理水库沉积物方面是有效的，但 SHPL 还为三台 50 兆瓦的混流式水轮机提供了额外的超音速火焰涂层防护，以防止任何颗粒进入电站进水口带来的不利影响。

分析:英语中的连词通常单独出现,而对应的汉语连词则需要成对出现。英译中时应注意增译组合出现的逻辑连词。本句中,"while"应译为"虽然······但是······"。

第五节　拓展延伸

一、中译英

（一）标准化管理情况

按照《××省人民政府办公厅关于全面推行水利工程标准化管理的意见》和《××省水利厅关于印发全面推行水利工程标准化管理实施方案的通知》精神,以及省水利厅关于××水利枢纽工程标准化管理"典型引路、示范先行"的试点工作要求,我局在2017年开展有关准备工作的基础上,于2018年全面推进××水利枢纽工程标准化管理创建工作,严格按照省厅"明确管理事项,确定管理标准,规范管理程序,科学定岗定员,建立激励机制,严格考核监督"六步法要求组织实施,全面实现了工程管理"管理责任明细化、管理工作制度化、管理人员专业化、管理范围界定化、管理运行安全化、管理经费预算化、管理活动日常化、管理过程信息化、管理环境美观化、管理考核规范化"标准化"十化"创建目标。

通过标准化管理工作创建,××水利枢纽建立了较为完善的工程标准化管理体系和运行管理机制,实现了工程管理事项清晰、岗位到人、责任到人、任务到人、监督到人,管理工作流程化、程序化、痕迹化及可追溯化,提高了工程管理信息化和自动化处理能力,提高了工程管理标准化、信息化水平,有力保障枢纽工程安全、持续、高效运行并充分发挥各项效益。

标准化管理创建后,××水利枢纽工程坝容坝貌和现场管理环境焕然一新,软硬件设施得到全面提升。特别是在编写管理手册和操作手册的过程中,管理局全体干部职工全程深入参与讨论修编,管理事项梳理清晰,岗位工作职责及工作流程熟悉掌握,并在实际工程管理中加以实践应用,成效良好。

××××年××月,水利部××副部长在调研××水利枢纽工程时充分肯定了工程的建设管理成就,认为工程建设成效显著,管理一流。××××年×

×月,省长××在调研××水利枢纽工程时指出,××水利枢纽工程建得好、管得好、取得的效益好。有关领导给予的高度评价,是对××水利枢纽工程标准化管理创建工作的充分肯定。

××××年××月,省水利厅考核验收组对××水利枢纽工程标准化管理创建工作进行了考核验收。枢纽工程高分通过水利工程标准化管理考核,达到一级标准,考核分居全省大中型水库(闸)类试点工程前列,在全省做出了示范和表率。

标准化管理信息化平台投入试运行后,根据实际情况,今年进行了一系列修改完善,组织了管理局及运行维养单位全体员工操作运用培训,通过平台操作、App操作试用,熟悉平台业务,提高工程管理工作效率。目前管理局各部门、运行维养单位正抓紧推进信息化平台运用的全面落实落地,提升标准化管理的信息化水平,努力全面实现工程标准化管理各项功能和成效。

(二)运行调度情况

××水利枢纽是一座具有防洪、发电、航运、灌溉等综合利用功能的大(1)型水利枢纽工程。××水利枢纽的运用调度遵循先考虑坝址上下游及大坝本身的防洪安全,再满足发电、航运、灌溉等用水要求的原则。

××水利枢纽防洪调度原则:依据上游来水流量,结合防洪限制水位(坝前水位),按"分界流量"采用"分级运行水位"的防洪、蓄水方式进行防洪和兴利调度。××××年确立的防洪限制水位(坝前水位):主汛期(4—6月)防洪限制水位为43.0米至43.5米(坝前控制运行水位),后汛期及非汛期防洪限制水位为44.0米至44.5米(坝前控制运行水位)。实际汛期与非汛期的划分按省防指确定的时限执行。

××水利枢纽兴利调度原则:确保水库大坝工程安全和上下游防洪安全;坚持"灌溉、生态、通航用水优于发电用水"的原则,以供定需,统一调度,分级管理;坚持计划用水、节约用水、科学用水,以保障水资源的合理利用;在满足灌溉、生态、通航用水及发电用水要求的前提下,水库尽量多蓄水以备干旱年份使用。

××水利枢纽生态供水调度方式:下泄生态流量主要通过发电机组尾水下泄,工程主体设计已考虑生态流量的下泄要求。若发电机组检修,生态流量则主要通过18孔泄水闸下泄。10月至次年3月以基荷发电方式下泄不得小于

221 米³/秒的流量,保证下游航道通航以及生产、生活、生态用水需求。当发电机组停机时,通过大坝泄水闸补足下泄流量至 221 米³/秒。4 月至 6 月,下泄生态流量不得小于 1200 米³/秒。当机组发电下泄水量不能满足要求时,通过大坝泄水闸补足下泄流量至 1200 米³/秒。7 月至 9 月,下泄生态流量不能小于 475 米³/秒。单台机组保证机组负荷率为 90% 时,可满足最小下泄流量要求;当单台发电机组负荷低于 90% 时,通过泄水闸补足下泄流量至 475 米³/秒。

××水利枢纽航运调度方式:当××坝址流量为 221～17400 米³/秒,且坝前水位在 42.70 米和 46.00 米之间、坝下水位在 30.30 米和 44.10 米之间时,××船闸按船只过往闸坝的需要正常通航。当××坝址流量小于 221 米³/秒或大于 17400 米³/秒时,××船闸停止通航。当××坝前水位低于 42.70 米或高于 46.00 米时,××船闸停止通航。当××坝下水位低于 30.30 米或高于 44.10 米时,××船闸停止通航。当××水利枢纽所在区域下大雨或暴雨,船闸引航道左岸冲沟流量(横流)较大时,××船闸停止通航。

投入运行以来,历年超过 10000 米³/秒的入库洪水有 8 次,以×××年××月××日的 16500 米³/秒为最大,且库区樟山、金滩、槎滩、柘塘防护区堤防超过设计洪水位。经过科学调度,历次洪水都顺利通过大坝,确保了工程安全和人民群众生命财产安全,发挥了重要的防洪功能和巨大的防洪效益。

二、英译中

Patrind HPP

The Patrind hydropower project features a weir site on the Kunhar River in Pakistan 120 km away from Islamabad towards the Northeast. The central block with a 44.0 m high gravity dam features two underflow and two overflow spillway bays for the diversion of high discharges.

The concept of sluicing high sediment loads, which are typical for the Himalayan region, is used as an approach for the sediment management. Thereby, the pool directly in front of the power intake, which is separated from the upstream reservoir by an overtopped cofferdam, is used as a natural settling pool.

The low flow velocities in the settling pool promote the deposition and storage of fine sediments up to 0.2 mm before reaching the power intake, where they could create abrasion during turbine operation. The sediment management concept schedules the flushing out of fine sediments through the underflow spillway once a year.

Additionally, a bypass tunnel of approximately 175 m length with an archway profile is implemented just upstream of the cofferdam in order to divert discharges higher than the maximal power plant discharge of 154 m³/s as well as flush sediments to the downstream river reach. The scheme of the natural settling pool for trapping fine sediment and the bypass tunnel for diverting all sediments is referred to as "Optimal Hybrid De-Sander System (OHDS)".

The VAW was commissioned by Saman Corp. In August 2014, Saman Corp. conducted with the hydraulic and sediment model investigations on the Patrind hydropower project with a physical model at a scale of 1:45. The model covers the gravity dam including the central block with two overflow spillway gates with integrated flap gates as well as two underflow spillway gates, the stilling basin, the power intake, the upstream cofferdam, the natural settling pool and the bypass tunnel. Upstream of the cofferdam, 300 m of the reservoir are modelled, while downstream of the weir, about 330 m of the river reach are reproduced.

The model investigations show that grains of 0.2 mm settle in the reservoir before they reach the power intake as long as a sufficient reservoir storage capacity can be maintained. In order to guarantee the long-term sustainability of the reservoir capacity, an appropriate operation regime is worked out, which includes the drawdown of the reservoir and the sluicing of large sediment loads during floods.

第三章　工程招投标文件

　　招标文件是招标工程建设的大纲,是建设单位实施工程建设的工作依据,是向投标单位提供参加投标所需要的一切情况。因此,招标文件的编制质量和深度,关系着整个招标工作的成败。建筑工程投标文件指具备承担建筑工程招标项目的能力的投标人,按照建设单位招标文件的要求编制的文件。

第一节　背景分析

一、招标文件

　　招标文件的繁简程度,要视招标工程项目的性质和规模而定。建设项目复杂、规模庞大的,招标文件要力求精练、准确、清楚;建设项目简单、规模小的,招标文件可以从简,但要把主要问题交代清楚。招标文件内容,应根据招标方式和范围的不同而有所变化。工程项目全过程总招标,勘察设计、设备材料供应和施工分别招标,其特点、性质截然不同,应从实际需要出发,分别提出不同的内容要求。

　　招标文件按照功能可以分成三部分:

　　一是招标公告或投标邀请书、投标人须知、评标办法、投标文件格式等,主要阐述招标项目需求概况和招标投标活动规则,对参与项目招标投标活动的各方均有约束力,但一般不构成合同文件。

　　二是工程量清单、设计图纸、技术标准和要求、合同条款等,全面描述招标项目需求,既是招标活动的主要依据,也是合同文件构成的重要内容,对招标人和中标人具有约束力。

　　三是参考资料,供投标人了解分析与招标项目相关的参考信息,如项目地址、水文、地质、气象、交通等参考资料。

　　招标文件至少应包括以下内容:

1. 招标公告。

2. 投标人须知,即具体制定的投标规则,使投标商在投标时有所遵循。投标须知的主要内容包括:(1)资金来源。(2)没有进行资格预审的,要提出投标商的资格要求。(3)货物原产地要求。(4)招标文件和投标文件的澄清程序。(5)投标文件的内容要求。(6)投标语言。尤其是国际性招标,由于参与竞标的供应商来自世界各地,因此招标文件必须对投标语言做出规定。(7)投标价格和货币规定。招标文件应对投标报价的范围做出规定,即报价应包括哪些方面,统一报价口径便于评标时计算和比较最低评标价。(8)修改和撤销投标的规定。(9)标书格式和投标保证金的要求。(10)评标的标准和程序。(11)国内优惠的规定。(12)投标程序。(13)投标有效期。(14)投标截止日期。(15)开标的时间、地点等。(16)品牌要求等。

二、投标文件

投标文件一般包含三部分,即商务部分、价格部分、技术部分。

商务部分包括公司资质、公司情况介绍等一系列内容,同时也包括招标文件要求提供的其他文件等相关内容,如公司的业绩、各种证件和报告等。

技术部分包括工程的描述、设计和施工方案等技术方案,工程量清单、人员配置、图纸、表格等和技术相关的资料。

价格部分包括投标报价说明、投标总价、主要材料价格表等。

一份完整的投标书应该包含以下内容:

1. 招标邀请函。招标邀请函是由招标人向投标人发出的邀请文件。单位收到招标邀请函,说明在资质、能力、技术方面满足招标项目的执行要求。

2. 投标人须知。这部分的内容一般在招标文件前面可以看到,主要告知投标人需要注意的问题、具体的招标程序和招标项目需要提供的材料。

3. 招标项目的技术要求和附件。只有满足招标项目要求的单位才可以投标,投标人需要提供自己的技术说明,并且要提供技术证明文件才能证明自身具有技术能力。

4. 投标书格式。每个单位的投标书格式都不一样,但是基本的投标程序、承担的义务和责任、报价等都要注明。

5. 投标保证文件。只有拥有投标保证文件,投标人才能证明已经向招标人提交了保证金,才有参与竞标的资格。

6.合同条件。合同条件关乎招投标双方的经济关系,只有中标单位才可以谈合同条件。

7.技术说明。要将自己的优势技术和技术效益讲明。

8.投标单位的基本资料,如营业执照、投标人身份证复印件、其他授权书等。

三、招投标的重要性

招标投标是一种国际上普遍运用的、有组织的市场交易行为,是贸易中的一种工程、货物、服务的买卖方式。

《中华人民共和国招标投标法》规定,在中华人民共和国境内进行下列工程建设项目包括项目的勘察、设计、施工、监理以及与工程建设有关的重要设备、材料等的采购,必须进行招标:(一)大型基础设施、公用事业等关系社会公共利益、公众安全的项目;(二)全部或者部分使用国有资金投资或者国家融资的项目;(三)使用国际组织或者外国政府贷款、援助资金的项目。前款所列项目的具体范围和规模标准,由国务院发展计划部门会同国务院有关部门制订,报国务院批准。法律或者国务院对必须进行招标的其他项目的范围有规定的,依照其规定。

建设工程招标投标具有以下重要意义:

1.通过招标投标提高经济效益和社会效益

我国社会主义市场经济的基本特点是要充分发挥竞争机制的作用,使市场主体在平等条件下公平竞争,优胜劣汰,从而实现资源的优化配置。

招标投标是市场竞争的一种重要方式,最大的优点就是能够充分体现"公开、公平、公正"的市场竞争原则,通过招标采购,让众多投标人进行公平竞争,以最低或较低的价格获得最优的货物、工程或服务,从而达到提高经济效益和社会效益、提高招标项目质量、提高国有资金使用效率、推动投融资管理体制和各行业管理体制改革的目的。

2.通过招标投标提升企业竞争力

通过招标投标可以促进企业转变经营机制,提高企业的创新活力,积极引进先进技术和管理经验,提高企业生产、服务的质量和效率,不断提升企业的市场信誉和竞争力。

3.通过招标投标健全市场经济体系

通过招标投标可以维护和规范市场竞争秩序，保护当事人的合法权益，提高市场交易的公平度、满意度和可信度，促进社会和企业的法治、信用建设，促进政府转变职能，提高行政效率，建立健全现代市场经济体系。

4.通过招标投标打击贪污腐败

招标投标有利于保护国家和社会公共利益，保障合理、有效使用国有资金和其他公共资金，防止其浪费和流失，构建从源头预防腐败交易的社会监督制约体系。在世界各国的公共采购制度建设初期，招标投标制度由于其程序的规范性和公开性，往往能对打击贪污腐败起到立竿见影的效果。

四、文本特点

招投标文件属于具有法律效力的商务合同，因此具有与一般法律文本相同的基本语言特征，如措辞准确、结构恰当、术语专业、思想缜密、文体正式、语义清晰等。

由于招投标文件的翻译属于法律文本的翻译，译者没有必要进行不必要的解释和艺术再创造。在翻译这类文本时，译者几乎没有任何自由发挥的空间。法律文本是用严谨的语言来编写的，翻译时不能对文本造成破坏。有时为了避免法律遗留的漏洞，语言的流畅性也会被忽略。为了突出其客观性和准确性，避免在后续介绍包含施工组织设计全部内容的施工项目和履行合同时产生歧义，翻译人员应了解招投标文件的语言特点。招投标文件在词汇方面有明显的特点，如频繁使用情态动词、专业术语和短语。为了使意思完整、表达严谨，避免误解，招投标文件中经常使用陈述句和祈使句，特别是并列句和复合句，以及被动语态来解释、说明投标各方规定的权利和义务。

由于源文本涉及专业技术知识，因此必定涉及专业术语和专用短语。鉴于招标文件的法律性质，招标文件在词汇、句法和语境层面上都具有独特的文体特征。在翻译过程中要特别注意遵循严格的翻译标准，避免出现歧义和含糊不清的表达。

五、译前准备

鉴于招投标文件的性质，翻译人员必须做好翻译前的准备工作。首先，译者应该阅读和学习原文的含义,了解原文的特殊风格和语言特征。其次,制作一个术语表。由于这是典型的技术性招投标文件,涉及大量的工程建设术语和行业知识,译者应向专业人士咨询,收集大量的资料。再次,确认所使用的翻译辅助工具。最后,制定翻译进度表,对译文进行校核。

第二节　专业术语

序号	中文	英文
1	安全生产许可证	safety production license
2	案例研究	case studies
3	板底筋	slab bottom bar
4	板筋	slab bar
5	板面筋	slab surface bar
6	饱满度	saturation
7	泵送剂	pump delivery agent
8	边筋	side bar
9	标底	base price limit on bids
10	标准和政策	standards and policies
11	玻璃工程	glazing works
12	不可抗力	force majeure
13	插筋	inserted bar
14	掺合料	mixture
15	撤标	withdrawal of bid
16	承包商	contractor
17	打桩工程	piling works
18	单步双信封	single-stage two-envelope
19	地下水位	water table
20	吊顶	ceiling
21	调直	straightness adjustment

续表

序号	中文	英文
22	顶撑	top strut
23	定距框	frame with fixed interval
24	定位钢筋	spacer bar
25	定位线	locating line
26	定型钢模板	fixed steel formwork
27	独立法人资格	independent legal personality
28	对拉螺栓	tie bolts
29	对账单	bank statement
30	筏板基础	raft foundations
31	法兰盘	ring flange
32	防水工程	waterproofing works
33	分包商	subcontractor
34	粉化	powdering
35	负筋	negative reinforcement
36	附加值	added value
37	钢结构工程	structural steel works
38	钢筋	steel reinforcement bar
39	隔板	bulkheads
40	给排水	plumbing and drainage
41	工程发包价	project contract issuing price
42	工程量清单	bill of quantities
43	工程事务协调员	engineering services coordinator
44	工程预算	project budget
45	工序	working procedure
46	管卡	pipe clamp
47	归方	de-squaring
48	合同承包价	contract price
49	合同条款	contract term

续表

序号	中文	英文
50	花岗岩石层	graphite stratum
51	环保专员	environmental officer
52	灰饼	mortar cake
53	汇款凭证	remittance voucher
54	减水剂	water-reducing agent
55	剪刀撑	herringbone strutting
56	建筑面积	gross building area
57	健康及安全专员	health and safety officer
58	胶粘剂	adhesive sizing agent
59	角钢	angle steel
60	金属工程	metal works
61	景观	landscape
62	聚氨酯	polyurethane
63	开标	open bid
64	开工日期	commence date
65	空鼓开裂	bulging and cracking
66	控制轴线	control axis
67	扣件	fastener
68	拉结筋	tie bar
69	立面卷材	facade roll
70	梁底垫块	cushion block
71	梁筋	beam bar
72	料斗	hopper
73	留槎	trough setting
74	楼地面工程	flooring finishing works
75	铝合金幕墙	aluminum curtain walling
76	履约保证金	performance security
77	马凳筋	barricade bar

续表

序号	中文	英文
78	抹灰工程	plastering works
79	木枋	woodwork
80	木作工程	joinery works
81	南部非洲发展共同体	Southern African Development Community
82	内撑筋	supporting bar
83	内撑条	internal strut
84	泥子	putty
85	起皮	scaling
86	砌体工程	masonry works
87	裙楼	podium building
88	日历天	calendar day
89	施工图纸	builders work drawing/construction drawing
90	室外工程	external works
91	水利水电工程施工总承包三级	Water Conservancy and Hydropower Engineering Construction General Contracting Level 3
92	水泥浆	cement mortar
93	松香水	mineral spirits
94	塔楼	tower
95	塔楼斜肋构架单元	diagrid units to the tower facade
96	套割	sleeved cutting
97	梯子筋	stair bar
98	替代方案	alternative offers
99	跳板夹道	springboard barricade for walking
100	通线	throughout line
101	砼块	concrete block
102	投标保证金	bid security
103	投标文件	bid proposal
104	土方工程	earth works
105	托油盘	oil tray

续表

序号	中文	英文
106	外立面	shop fronts facades
107	外加剂	additive
108	五金	ironmongery
109	物资采购经理	material resource and procurement manager
110	现场踏勘	site survey
111	现浇带	cast-in-place belt
112	限位钢筋	limit bars
113	香蕉水	lacquer thinner
114	响应性	responsiveness
115	絮凝剂	flocculating agent
116	压光	press and polish
117	引气剂	air-entraining agent
118	油漆工程	paint works
119	釉面砖	glazed tile
122	预制混凝土工程	precast concrete works
121	员工简历	staff CVs
122	招/投标	bid/tender
123	招标文件	bid invitation/proposal request
124	找方	squaring
125	执行/管理总结	executive/management summary
126	质量控制经理	quality control manager
127	中级承包商	intermediate contractor
128	主筋	major bar
129	柱立筋	vertical bar of the column
130	柱主筋	major bars for the column
131	转包/分包	subcontract
132	桩基础	pile foundations
133	资金来源	source of fund
134	资质信息	qualification information

第三节　翻译案例

案例一：中译英

原文：

水利水电工程小额工程施工

招标文件

招　标　编　号：<u>　闽建发招字〔20××〕×××号　</u>

项　目　名　称：<u>　××区防洪工程　　　　　　　　</u>

招　标　人：<u>　××区管委会　　　　　</u>（盖章）

法　定　代　表　人：<u>　　　　　　　　　　　　　　</u>（盖章）

招　标　代　理　单　位：<u>　××公司　　　　　　　　</u>（盖章）

法　定　代　表　人：<u>　　　　　　　　　　　　　　</u>（盖章）

日期：20××年××月

目　　录

第一部分　投标人须知

（一）总则

1. 工程说明

1.1　项目名称：××区防洪工程。

1.2　建设地点：××镇。

1.3　建设规模：桩号 0＋000——0＋×××。

1.4　承包方式：包工包料。

1.5　质量标准：合格。

1.6　招标范围：按图纸并结合工程量清单。

1.7　施工总工期：30 日历天。

1.8　资金来源：自筹。

2. 获取招标文件方式

投标人直接从××市招投标中心网站下载小额工程的招标信息。需要施工图纸的，直接向招标人（招标代理机构）联系购买。

3. 投标人的资格资质要求

3.1　投标人资质等级要求具备建设行政主管部门核发的水利水电工程施工总承包三级及以上（不分主、增项差别），并具有有效的安全生产许可证的企业。

3.2　凡在××市内注册登记，具有独立法人资格，符合规定资质条件且未因违法违规被限制参加相应项目投标的潜在投标人均可参与投标。

3.3　近三年内，未被本市县区级以上相关行政监督部门记入不良行为的。

4. 投标有效期 60 日历天（从投标截止之日算起）。

5. 投标费用

投标人应承担其编制投标文件与递交投标文件所涉及的一切费用，不管投标结果如何，招标人和招标代理机构对上述费用不负任何责任。

6. 本招标工程项目不允许违法转包、分包。除地基基础和主体工程外确需分包的，应符合有关法律、法规规定并经招标人同意。

7. 本文件依据市政府×政综〔20××〕×××号文制定。招投标活动的有关问题处理按照市政府×政综〔20××〕×××号文件执行。

（二）招标文件

8. 招标文件

招标文件由投标人须知、合同条款、投标文件组成(格式)、工程量清单、工程预算、施工图纸、修改和澄清的答疑文件组成。

9. 现场踏勘

9.1　招标人不组织现场勘察。招标人向投标人提供的有关现场的数据和资料，是招标人现有的能被投标人利用的资料。招标人对投标人做出的任何推论、理解和结论均不负责任。

9.2　参加现场踏勘等咨询活动不是强制性的，由投标人自行决定。

9.3　投标人可为踏勘目的进入招标人的工程项目现场，但投标人不得因此使招标人承担有关的责任和蒙受损失。投标人应承担现场踏勘的费用、责任和风险。

10. 招标人不组织投标答疑会

11. 招标文件的澄清及修改

投标人对工程情况和招标文件(含工程量清单及工程预算价)有疑问的可于×××年××月××日 17:00 前直接登录××市招投标中心网向招标人提出(应列明招标项目名称和编号)。招标人将于×××年××月××日 17:00 前在××市招投标中心网上公布澄清及修改的答疑内容。

12. 工程预算价×××元(人民币)

13. 工程发包价

工程发包价=工程预算价×(1−下浮率)，为×××元(人民币)。本工程下浮率为<u>12%</u>。

14. 合同承包价

招投标双方应在签订施工合同时，对工程发包价中的劳保费用进行调整，调整后的价格即为合同承包价。

15. 投标保证金的缴交和退还

15.1　投标保证金金额为人民币：<u>壹万元整</u>(￥<u>10000</u>元)，投标人必须按规定交纳。

银行账户名称：×××。

开户行：建设银行×××支行。

账　号：×××××××××××××××××××××。

15.2　投标保证金手续须在截标前自行办妥。投标保证金应当转入招标人指定的银行专户,并写明投标人、投标项目名称或招标项目编号。缴交时间确认以招标文件指定银行在截标时出具的对账单为准。截标时,市招投标中心应当负责查询该项目投标保证金入账情况,并及时通知招标人。

15.3　投标人应按招标文件规定的金额,将投标保证金从投标人银行基本账户按招标项目单标单笔一次性汇入招标文件指定的银行账户。若因投标人分批分笔缴交投标保证金而导致投标保证金统计失误的,由投标人自行负责。

15.4　投标人在缴纳投标保证金时,须注明招标项目编号(即招标文件上注明的招标编号)或招标项目名称。若招标文件指定银行出具的对账单或投标人汇款凭证上有体现招标项目编号或招标项目名称,则视为该投标人已提交投标保证金。

15.5　开标完成后,未中标的投标人的保证金次日退还至其基本账户,中标人的投标保证金在签订施工合同后5日内退还至其基本账户。

15.6　未按要求提交投标保证金的投标将被视为无效投标。

15.7　发生下列情况之一,投标保证金将被没收:

(1)开标后在投标有效期间,投标人撤回其投标书的;

(2)中标人放弃中标的或不按规定签约的;

(3)中标人在规定的期限内未提交履约保证金的。

译文：

Construction of Small Water Conservancy and Hydropower Projects
Bid Invitation

Invitation for Bids No. :　Minjianfa Zhaozi〔20 × ×〕× × ×

Project Name:　Flood Control Project of × × Zone

Bid Inviter:　× × Zone Management Committee　(Seal)

Legal Representative:　(Seal)

Bidding Agent:　× × Co. , Ltd.　(Seal)

Legal Representative:　(Seal)

Date: × ×

Content

Part 1   Instructions to Bidders

（1）General Provisions

1. Engineering Specification

1.1 Project name: × × Zone Flood Control Project.

1.2 Construction sites: × × Town.

1.3 Construction scale: pile number 0 + 000——0 + × × ×.

1.4 Contracting method: contract labor and materials.

1.5 Quality standard: qualified.

1.6 Scope of bidding: according to the drawing and combined with the bill of quantities.

1.7 Total construction period: <u>30</u> calendar days.

1.8 Source of funds: self-financing.

2. Ways to Obtain Bid Invitation Documents.

The bidder directly downloads the bid invitation information of small projects from the website of × × City Bid Center, and if construction drawings are needed, contacts the bid inviter（bidding agency）for purchase.

3. Qualification Requirements of Bidders

3. 1 The bidder's qualification level requires the enterprise with Water Conservancy and Hydropower Engineering Construction General Contracting Level 3 or above issued by the construction administrative department (regardless of the difference between the main and additional items) and valid safety production license.

3. 2 Any potential bidder who is registered in × × City, has an independent legal personality, meets the prescribed qualification conditions and is not restricted from participating in the corresponding project bidding due to violations of laws and regulations may participate in the bidding.

3. 3 In the past three years, the bidder has not been recorded as bad behavior by the relevant administrative supervision departments at or above the county and district level of the city.

4. The bid is valid for 60 calendar days (counting from the bid closing date) .

5. Bidding Fee

The bidder shall bear all expenses involved in the preparation and submission of bid submission documents, regardless of the result of the bid, the bid inviter and the bidding agency shall not be liable for the above expenses.

6. Illegal subcontracting are not allowed in this bidding project. If subcontracting is necessary except for foundation and main works, it shall comply with relevant laws and regulations and be approved by the bid inviter.

7. This document is formulated in accordance with X. Z. Z Document〔20 × × 〕 No. × × × of × × Municipal Government. The handling of issues related to the bidding activities shall be carried out in accordance with the X. Z. Z Document〔20 × × 〕No. × × × of × × Municipal Government.

(2) Bid Invitation Documents

8. Bid Invitation Documents

The bid invitation document consists of instructions to bidder, contract terms, bid proposal(format) , bill of quantities, project budget, construction drawings, modification and clarification of the Q&A documents.

9. Site Survey

9. 1 The bid inviter shall not organize site survey. The relevant data and materi-

als of site provided by the bid inviter to the bidder are the existing materials that can be used by the bidder, and the bid inviter shall not be responsible for any inference, understanding and conclusion made by the bidder.

9.2 Participation in consulting activities such as site survey is not mandatory and shall be determined by the bidder.

9.3 The bidder may enter the project site of the bid inviter for the purpose of survey, but the bidder shall not make the bid inviter bear relevant responsibilities and suffer losses. The bidder shall assume the costs, responsibilities and risks of site survey.

10. The bid inviter shall not organize a bidding question-and-answer meeting.

11. Clarification and Modification of Bidding Documents

If the bidder has any questions about the project situation and the bid invitation documents (including the bill of quantities and the project budget price), it may submit to the bid inviter through × × City Bidding Center website before 17:00 on date × × (the name and number of the bidding project should be specified). The bid inviter will publish the clarification and modification contents on the × × City Bidding Center website before 17:00 on date × ×.

12. Project budget is × × × yuan (RMB).

13. Project Contract Issuing Price

Project contract issuing price = project budget price × (1 − float rate), is × × × yuan (RMB). The floating rate of this project is 12%.

14. Contract Price

When signing the construction contract, both parties shall adjust the labor insurance fee in the project contract issuing price, and the adjusted price shall be the contract price.

15. Payment and Return of Bid Security

15.1 The bid security shall be RMB: 壹万 (Chinese numeral capital case) Yuan (￥10000 Yuan), which shall be paid by the bidder in accordance with the regulations.

Bank account name: × × ×.

Opening bank: CCB × × × Branch Bank.

Account number: ×.

15.2 Bid security procedure shall be completed by bidder before the bid closing. The bid security shall be transferred to the designated bank account of bid inviter, and the name of the bidder and the bidding project, or the number of the bidding project shall be clearly indicated. The confirmation of the payment time shall be subject to the statement issued by the bank designated in the bid invitation documents at the time of bid closing. When the bid is closed, the Municipal Bidding Center shall be responsible for inquiring the entry of the project bid security and notify the bid inviter in time.

15.3 The bidder shall, according to the amount specified in the bid invitation documents, transfer the bid security from the basic account of the bidder's bank to the designated bank account in the bid invitation documents in one lump sum according to the single bid project. The bidder shall be responsible for any statistical error of bid security caused by the payment in batches.

15.4 When paying the bid security, the bidder shall indicate the bidding project number (the bidding number indicated on the bid invitation document) or name. If the statement issued by the bank designated in the bidding document or the bidder's remittance voucher shows the bidding project number or name, the bidder shall be deemed to have submitted the bid security.

15.5 After the completion of bid opening, the bid security of the failed bidder shall be returned to its basic account the next day, and the bid security of the winning bidder shall be returned to its basic account within five days after the signing of the construction contract.

15.6 A bid that fails to submit a bid security as required will be deemed invalid.

15.7 The bid security will be forfeited under any of the following conditions:

(1) The bidder withdraws its bid submission during the validity period of the bid after the bid opening;

(2) The winning bidder gives up the bid or fails to sign the contract as re-

quired;

(3) The winning bidder fails to submit the performance security within the pre-scribed time limit.

案例二:英译中

原文:

× × Buildings
Technical Proposal

Contents: <u>Technical bid (construction organization design)</u>

Bidder: _____ (official seal)

Legal representative: _____

Or its authorized agent: _____ (signature or seal)

Chapter One   Project Profile and Overview

1.1 Project Profile

Project name: × × Buildings.

Construction period: 24 months.

Project contents and time requirements: the project is mainly composed of a tower and three podium buildings with their corresponding outdoor works. The area of each part is as follows.

Building No.	Gross Building Area	Storey
Building 1A	× × × m^2	7 storeys including 3-level basement
Building 1B	× × × m^2	7 storeys including 3-level basement
Building 2	× × × m^2	17 storeys including 2-level basement
Building 3	× × × m^2	5 storeys including 2-level basement
Total	× × × m^2	

Commencement date: October 3, 2023.

Specific construction works:

(1) Earth Works; (2) Piling Works (105 No's, Diameter: 450 mm, Average Depth: 5 m) ; (3) Concrete Works; (4) Precast Concrete Works; (5) Masonry Works; (6) Waterproofing Works; (7) Joinery Works; (8) Flooring Finishing Works; (9) Structural Steel Works; (10) Metal Works; (11) Plastering Works; (12) Glazing Works; (13) Paint Works; (14) External Works.

The subcontracted items:

Electrical Installation; Mechanical Installation; Lift Installation; Fire Services Installation; Wet Services Installation; Aluminum Curtain Walling, Shop Fronts and Facades to the Tower; Diagrid Units to the Tower Facade; Aluminium Cladding to Diagrid Edge Columns and Beams; Aluminium Curtain Walling, Shop Fronts and Facades for Buildings 1A, 1B and 3; Landscaping, Irrigation and Roof Gardens; Ceilings and Bulkheads; Ironmongery; Plumbing and Drainage.

Geological condition:

(1) The water table is variable and the highest point is 2. 8 meters underground.

(2) It is concluded that the hard graphite stratum is located from a depth of 2. 1 meters underground and the deepest point is 11. 2 meters underground.

(3) The geological report suggests pile foundations for podium Building 1A and pile foundations or raft foundations for Building 2.

1. 2 Overview

Management, programming/planning, coordination and resourcing:

In consideration of the specific characteristics of this project regarding management, programming/planning, coordination and resourcing for the execution of the works, × × Company will:

Arrange all the works in a proper and organized sequence in accordance with the construction plan, concentrate on the subcontract works and pay attention to the important points in the project.

Optimize production and reduce the completion time through professional subcontracting of project and management so as to meet the interests of employers and ×× Company.

All of the above requirements for submission and approval will be incorporated into our project construction plan to ensure timely and orderly conduct of the project.

As main contractor, we shall mobilize and set up our project management team and workforce, together with all subcontractors after on-site meeting for the proper organization, management, coordination and project control. All members of the project management team have their own responsibilities, clear division of duties thus forming a cohesive group. The planning and control of the project should be organically combined so that all participants in the project have decision-making power.

A qualified health and safety officer, an environmental officer and a quality control manager will also be employed for the full duration of the contract to ensure the project can be completed on time and within the budget constraints laid down by the employer. All consultants, employer's representatives and main contractor representatives shall be present at progress meetings held on site. General contractual matters will be discussed at the progress meeting. Intermediate contractor/subcontractors on-site meetings will be held on a fortnightly basis to discuss and resolve drawing and technical related and coordination issues between subcontractors and main contractor. The above meetings will be chaired by our engineering services coordinator. Copies of the record of these meetings will be distributed to all consultants, employer's representatives and subcontractors. Relevant consultants will conduct on-site inspections as required. All shop drawings, installation and builders work drawings, technical literature, specifications and samples should be in compliance with the project construction program. All the members of the construction site shall form a working group to clarify their respective responsibilities under the leadership of the project manager and formulate the project control plan so as to fully participate in the decision-making process of all the participating parties of the project.

Through the use of project management software and experienced and qualified staff, ×× Company is able to implement project construction plan that enables the company, consultants and the employer to switch directly to the site of the project and monitor actual progress in an accurate, predictive and comprehensive manner. The detailed information included in the project construction program also allows the employer, through inspection and observation, to accurately monitor current status, the overall and detailed progress of the project at any time during the construction period. Monthly progress update reports and cash flow updates will be available for verification by the principal agent. In order to promote the development of the project, the staff on site and in the headquarter will assist the management team to handle and perform specific tasks, reduce the low efficient work due to the sick or injured and finish the construction project on time.

Our resource strategy may be divided into three sections as follows:

1) Labor: We intend to employ people with experience in the construction industry for the senior management positions. The contractor can be recruited locally and the construction workers generally come from our current employees and if required, they can be sourced abroad. The resource strategy for labor will be compiled under the management of HR manager and will briefly consist of:

- Locally available labor sourcing
- Abroad available sourcing (if necessary)
- Work permits
- Botswana Labor Act requirements
- Contracts
- On-site labor management

All special subcontractors labor resourcing requirements will be monitored by the main contractor.

2) Material: Locally produced materials are procured through local suppliers, and material samples will be sent to professional institutions for material testing in appropriate cases. Imports from the Southern African Development Community would be made if there is no provision for the use of locally produced materials. The resource strategy for materials will be compiled by the material resource and procure-

ment manager and will briefly consist of:

- · Locally available material sourcing
- · Abroad available material sourcing
- · Purchasing
- · Expediting supplier quality inspections (where possible)
- · Logistics/Import duties
- · Transport
- · On-site material delivery management and recording

All subcontractors material resourcing requirements will be monitored by the main Contractor.

3) Plant: We will use our own factory in the possible situation and adopt professional equipment as required. × × Company will make prior arrangements with alternative suppliers and/or personnel agencies as required. When there exists apparent demand of plant and labor, × × Company will first turn to the consultants.

<h3 style="text-align:center">Chapter Two   Project Quality Control Measures</h3>

2.1 Quality Control Measures in Key Working Procedures

2.1.1 Quality Control Measures for Steel Reinforcement Bars

Item	Quality Control Points/Items	Quality Control Measures
1	Steel reinforcing bar fabrication	1. Engineers will be randomly dispatched to check precision of the fabricated steel reinforcement bars at the suppliers workshop as per the drawings to ensure the dimensions are accurate. 2. The steel reinforcing bar fabrication process will be strictly controlled. Straightening, cutting, bending, molding and other procedures will be processed by machinery equipment. 3. After the steel bars are transported to construction site, the steel fixers will be in charge of the management to ensure that the steel reinforcement bars are sorted, kept and used in accordance with the applied parts and time.

续表

Item	Quality Control Points/Items	Quality Control Measures
2	Steel reinforcement bar locating and protection layer	1. A frame with fixed interval is used to control the interval between major bars for the columns. The frame will be placed before pouring concrete slab at the height of 30 – 50 cm above the top of the slab and firmly bound with the vertical steel reinforcement bars for the column. After the concrete is poured, the frame will be disassembled for circulating use before the vertical steel reinforcement bars are bound with the upper layer. 2. The beam bars are concerned with falling of the 2nd row of negative moment reinforcement and uneven thickness of the protection layer for side beam. Therefore, if the upper and lower major bars for the beam are two or three rows, the short bars of $\Phi 25 @ 1000$ mm centers will be placed between rows along the beam to separate the rows. Length of the short bars equals beam width minus 2 times of the thickness of the protection layer. Cushion block at the bottom of the beam will be placed alternatively, and the bars on both sides of the beam will be added with plastic circular ring cushion block in the pattern of plum blossom. 3. As for the slab bars, the concern is the falling of the negative moment of bars. The barricade bar will be used to prevent the falling, the bars shall not be stepped after binding. Therefore, in the process of binding, springboard barricade for walking will be provided until the concrete pouring is finished. 4. Except the sloe plate of bottom slab, the bar protection layer for the other sections will be mainly ensured by using the plastic finished cushion block.
3	Preset holes for electrical and mechanical installation	It will be specified that the bars for wall and floor slab opening must be completed once if its diameter is more than 200 mm, and cutting after binding is not allowed. The organization personnel shall use the computer CAD technology to draw the structure reservation, the preset hole and refine the reinforcement to eliminate the phenomenon that the water and electricity are reserved at random in the reinforcement project, so as to ensure the accurate location and safe structures of the embedded hydro-power projects.

2.1.2 Quality Control Measures for Formwork

Item	Quality control points/items	Quality Control Measures
1	Dimensions of the column	1. Set out the control line. 2. Use the stair bar and internal strut. 3. Use the tie bolts.
2	Preset openings for door and window	1. If the width is over 1.5 m, fixed steel formwork will be used. 2. If the width is less 1.5 m, wooden formwork will be used. The corner will be formed by 10 mm × 10 mm angle steel and M16 bolts. 3. Sponge strip shall be added around the formwork. 4. Limit bars shall be added on both sides of the formwork and spacer bars on the bottom to control the size of the opening. 5. The concrete shall be poured from both sides of the opening simultaneously.
3	Prevention of the broken of the column root	1. When pouring the top concrete slab, the area within 15 cm from the side line of the column will be pressed and polished, the excessive floating mortar will be removed. 2. Mortar or sponge strip will be used to seal the root of the column formwork.
4	Prevention of formwork expansion and deviation	1. Reinforce the strength and toughness control when designing the formwork. 2. Set out the location line and control axis before supporting the formwork. 3. Use the stair bar, top strut and other locating measures.
5	Prevention of misalignment and mortar leakage	1. "wedge set" shall be used to connect the steel formwork. 2. Sponge strip shall be added in the jointing.
6	Installation and set-out	1. Survey the control axis network and formwork control line. 2. Set out the side line of the column and check the control line. 3. After vertical bar binding, the elevation control point will be marked on the top of major bars on each floor.

Item	Quality control points/items	Quality Control Measures
7	Flatness of the slab	1. The formwork supporting must be calculated. 2. The formwork must be selected before assembled. 3. Set out the control line when installing the formwork. 4. Supports of the upper and lower layers shall be in the same straight line, and the bottom of supports will be added with base or wood block.
8	Formwork at the construction joint	Drill on the formwork of the construction joint, and the slab bar will go through the formwork.
9	Joint treatment	1. Amplify the sample, and set out the control line. 2. Sponge strip will be added for the purpose of reinforcement.
10	Prevention of arching	1. When the span is equal to or more than 4 m, the arching height shall be $1/1000 - 3/1000$ the length of the full span. 2. Draw the throughout line.

2.1.3 Quality Control Measures for the Concrete Work

Item	Quality control points/items	Quality Control Measures
1	Treatment measures for the concrete construction joints	1. Vertical construction joints of wall: the dense double-layer steel mesh will be bound on the wall reinforced bar, and 50mm thick wood will be used to retain the concrete. After the wall formwork is removed, mark lines on both sides of the wall equally 50mm away from the construction joint, cut a straight joint 5mm deep along the ink line with the concrete cutting machine and remove the exposed gravel from the soft concrete layer out of the straight layer to ensure joint quality of the concrete. 2. Horizontal construction joints at the top of the wall: pouring the wall body concrete shall be $20 - 30$ mm higher than the top slab roof. After the wall formwork is dismantled mark the bottom line of the top slab, pop the bottom line, cut a straight joint 5 mm deep with cutter and remove the exposed gravel of the soft concrete layer above the straight joint.

续表

Item	Quality control points/items	Quality Control Measures
1	Treatment measures for the concrete construction joints	3. Construction joint on the bottom of the wall and column: remove the floating mortar and make it 10 mm concaved to ensure the quality of concrete joint, then fully dampen and flush but not to accumulate water. 4. Construction joints at the top slab: 15 mm thick wooden strips shall be placed under the construction joints at the top slab to ensure the protection layer of iron reinforcement. The net distance between upper and lower iron is ensured by wooden boards, and the sides of wood boards contacting the upper and lower iron are cut into gaps through the spacing of steel bars and stuck on reinforcement bars. 5. Concrete pouring at the construction joint: when pouring concrete continuously at construction joints, the compressive strength of concrete poured shall not be less than 1.2 N/mm^2. Before pouring the concrete, it is advisable to pave a 5 – 10 cm layer of cement mortar with the same mixing proportion at the construction joints. The concrete shall be compacted to be fully mixed.

2.1.4 Quality Measures for Waterproofing Work

Item	Quality control points/items	Quality Control Measures
1	Pavement with Coiling material	1. The base layer treatment agent is evenly coated, and after drying, it can be paved with coiling material. 2. Roof waterproofing coiled material shall undergo a trail pavement and leave sufficient material. The enhanced additional layer shall be paved first, then the plane coil to the corner and finally the facade roll from upper to bottom.
2	Waterproof for construction joint and expansion joint	Add an additional waterproof layer.
3	Waterproofing treatment at the corner	1. An additional layer of the same material with the waterproof layer will be added. 2. An additional coiled material (empty paving) will be added in the outer corner of the basement and the junction of the plane.

译文:

×××大楼项目技术标

投标文件

投标文件内容: <u>投标文件技术标(施工组织设计)</u>

投　标　人: _____ (盖公章)

法定代表人

或其委托代理人: _____ (签字或盖章)

第一章　工程概述

1.1　项目简介

◆项目名称:博茨瓦纳 Fairscape 区域商务中心大楼。

◆总工期:24 个月。

◆工程内容和时间要求:该项目主要由一栋塔楼和三栋裙楼及相应的室外工程组成。各部分建筑面积如下:

楼号	建筑面积	层数
Building 1A	×××m²	地上 4 层,地下 3 层
Building 1B	×××m²	地上 4 层,地下 3 层
Building 2	×××m²	地上 15 层,地下 2 层
Building 3	×××m²	地上 3 层,地下 2 层
合计	×××m²	

◆计划开工日期:2023 年 10 月 3 日。

◆具体施工工作内容

(1)土方工程;(2)打桩(105 根,直径 450 mm,平均深度 5 m);(3)混凝土工程;(4)预制混凝土工程;(5)砌体工程;(6)防水工程;(7)木作工程;(8)楼地面工程;(9)钢结构工程;(10)金属工程;(11)抹灰工程;(12)玻璃工程;(13)油漆工程;(14)室外工程。

◆业主指定分包的项目

电,机械,电梯,消防,水,铝合金幕墙,塔楼外立面,塔楼斜肋构架单元,通往斜肋构架的铝合金梯,裙楼 1A、1B 和 3 的铝合金幕墙和外立面,景观、灌溉及屋面花园,吊顶和隔板,五金,给排水等。

◆项目地质情况

(1)现场地下水位分布不均,最浅水位约 2.8 m。

(2)整个场地在最浅处 2.1 m 左右有坚硬的花岗岩石层;最深处在 11.2 m 左右有坚硬的花岗岩石层。

(3)地质报告建议 1 号裙楼采用桩基础,2 号塔楼采用桩基础或筏板基础。

1.2　项目概述

管理、规划/策划、协调及资源回收:

考虑到该项目的管理、规划、计划、协调和资源分配的具体特点,××公司将:

按照计划合理安排工程顺序,集中完成分包工程,同时注意工程中的重要节点。

通过项目和管理的专业分包,优化生产,尽量缩短完工时间,以满足雇主和××公司的利益。

上述所有关于提交和批准的规定将纳入我们的项目建设计划,以确保工程及时和有序进行。

作为主承包商,我们将在现场会议后,与所有分包商一起,动员和建立项目管理团队和员工队伍,以便更好地组织、管理、协调和控制项目。项目管理团队的所有成员各司其职,分工明确,形成高凝聚力的团体。项目的计划和控制应有机结合,使所有的项目参与者有决策权。

合同期内聘用合格的健康及安全专员、环保专员及质量控制经理,以确保该项目能按时并在雇主制定的预算限制内完成。所有顾问、雇主代表和总承包商代表应出席在现场举行的进度会议。一般的合同事项将在最近的进度会议上讨论。中级承包商/分包商现场会议将每两周举行一次,以讨论和解决分包商和主承包商之间的技术相关问题和协调问题。上述会议将由我们的工程事务协调员主持。这些会议记录的副本将分发给所有顾问、雇主代表和分包商。有关顾问会根据要求进行现场视察。所有的车间图纸、安装和施工图纸、技术文献、规格和样品应该符合项目建设计划的要求。施工现场的所有成员组成工作小组,在项目经理的领导下明确各自的工作职责,制订项目控制计划,使项目的所有参与方充分参与决策环节。

通过使用项目管理软件和经验丰富的员工,××公司能够实施项目建设计划。该计划使公司、顾问和雇主能够在准确、预测和全面的方式下,直接切换到项目现场以便跟踪和监控项目进展。项目建设计划中包含的详细信息也允许雇主通过检查和观察,在项目周期中随时准确地监控项目的现状、总体和详细的进展情况。每月的进度更新报告和现金流更新情况将由主要代理人员进行核实。为了推进项目发展,现场和总部的工作人员将协助管理团队管理和执行具体任务,减少因人员生病或受伤对生产力造成的影响,按时完成合同规定的建设项目。

我们的资源策略可分为以下三个部分:

1)劳动力:我们打算聘用有建筑行业经验的人员担任高级管理职位。包工头可以在当地招聘,建筑工人可部分采用现有员工,如有需要,也可以从当地招聘。人力资源战略将在人力资源管理经理的管理下编制,并大致包括:

- 本地可用的劳动力资源
- 国外资源(如果需要)
- 工作许可证
- 博茨瓦纳劳工法案的要求
- 合同
- 现场劳动力管理

所有专业分包商的劳动资源需求将由主承包商负责监督。

2)材料:当地生产的材料通过当地供应商采购,在适当的情况下,将材料样品送往专业机构进行检测。如果没有规定必须采用当地生产的材料,则从南部非洲发展共同体地区进口。材料资源战略将由物资采购经理编制,并大致包括:

- 本地可用的材料资源
- 国外材料资源
- 采购
- 加快供应商质量检查(在可能的情况下)

- 物流/进口关税
- 运输
- 现场物料交付管理和记录

所有专业分包商的材料再采购需求将由主承包商监控。

3）设施：我们将在可能的情况下使用自己的工厂，并根据需要采用专业设备。××公司将根据需要与其他供应商和/或人事机构做出事先安排，如果明显存在设施和劳动力的短缺，将首先咨询顾问的意见。

第二章 工程质量控制措施

2.1 关键工序质量控制措施

2.1.1 钢筋工程质量控制措施

序号	质量控制点/项目	质量控制措施
1	钢筋制作	1. 安排工程师不定时地到供应商处按照图纸说明检查钢筋制作的精度，确保钢筋配料尺寸准确。 2. 严格控制钢筋制作，调直、切断、弯曲、成型等工序均采用机械设备加工。 3. 钢筋加工完毕运至现场后设专人负责成型钢筋的管理工作，保证成型钢筋按使用部位、使用时间进行分类分批编组编号码放、保管及发放。
2	钢筋定位及保护层控制	1. 柱主筋间距采用定距框控制。定距框在板混凝土浇筑前放置，高出板顶 30 cm—50 cm，与柱的竖向钢筋绑扎牢固。混凝土浇筑完毕，上层柱竖向钢筋绑扎前将定距框拆卸，供周转使用。 2. 梁筋主要涉及二排筋的坠落和梁侧保护层厚度不均的问题。梁上下部主筋为二排或三排时，在排与排之间沿梁长方向设置 $\Phi25@1000$ 的短钢筋，将各排钢筋分开。短钢筋长 = 梁宽 − 2 倍的保护层厚度。梁底垫块交错布置，梁两侧钢筋上加塑料环圈垫块，按梅花形设置。 3. 板筋主要涉及负筋下坠的问题，除用马凳筋外，对现浇板钢筋更重要的是绑扎成型后不要踩踏。板筋绑扎的过程中，应设置供行走用的跳板马道，直至混凝土浇筑完成。 4. 除基础底板外，其他部位的钢筋保护层主要采用塑胶成品垫块。

续表

序号	质量控制点/项目	质量控制措施
3	机电安装预设孔	施工准备计划明确要求：大于 200 mm 的所有墙体和楼板洞口的钢筋必须一次配筋下料施工完成，不允许钢筋绑扎完成后再切割。组织人员利用计算机 CAD 技术绘制出结构预留、预设孔图，细化配筋，杜绝在钢筋工程中随意预留水电孔的现象，确保水电预留预埋位置准确和结构安全。

2.1.2　模板工程质量控制措施

序号	质量控制点/项目	质量控制措施
1	立柱尺寸	1. 测量控制线。 2. 使用梯子筋、内撑条。 3. 使用对拉螺栓。
2	预设门窗开口	1. 超过 1.5 m 宽的洞口采用定型钢模板。 2. 小于 1.5 m 宽的洞口采用木模板，角部由 10×10 的角钢及 M16 螺栓组成。 3. 在模板周边加设海绵条。 4. 在模板两侧加设限位钢筋，底部设定位钢筋，控制洞口尺寸。 5. 从洞口两侧同时浇筑混凝土。
3	柱烂根预控	1. 在顶板砼浇筑时，对柱边线 15 cm 内的范围进行压光，去除多余的表面浮浆。 2. 柱模根部采用砂浆或海绵条封堵。
4	胀模、偏位预控	1. 设计模板时加强强度和刚度控制。 2. 模板支设前放好定位线、控制轴线。 3. 使用梯子筋、顶撑等定位措施。
5	错台、漏浆预控	1. 钢模板连接采用"子母扣"形式。 2. 拼接处加设海绵条。
6	安装放线	1. 测量控制轴线网和模板控制线。 2. 放置柱边线和检查控制线。 3. 竖向钢筋绑扎完成后，在每层竖向主筋上部标出标高控制点。
7	板面平整	1. 模板支撑必须经过计算。 2. 模板拼装前，必须经过挑选。 3. 模板支设时，放出控制线。 4. 上下层支撑在同一直线上，支撑下加底座或垫木方。

续表

序号	质量控制点/项目	质量控制措施
8	施工缝处的模板	在施工缝处的模板上钻眼,板筋穿过模板。
9	梁、柱节点模板及梁板柱模拼缝处理	1. 放大样,放出控制线。 2. 加设海绵条,做好加固处理。
10	起拱预控	1. 跨度等于或大于 4 m 时,模板起拱高度宜为全跨长度的 1/1000—3/1000。 2. 拉通线。

2.1.3 混凝土工程质量控制措施

序号	质量控制点/项目	质量控制措施
1	混凝土施工缝处理措施	1. 墙体竖向施工缝:可用密目双层钢丝网绑扎在墙体钢筋上,外用 50 mm 厚木板封挡混凝土。当墙模拆除后,在距施工缝 50 mm 处的墙面上两侧均匀弹线,用砼切割机沿墨线切一道 5 mm 深的直缝,再用钎子将直缝以外的混凝土软弱层剔掉露石子,清理干净,保证混凝土接槎质量。 2. 墙体顶部水平施工缝:墙体混凝土浇筑时,应高于顶板底 20 mm—30 mm。墙体模板拆除后,弹出顶板底线,在墨线上 5 mm 处用切割机切割一道 5 mm 深的水平直缝,将直缝上的混凝土软弱层上的石子剔除并清理干净。 3. 墙、柱底部施工缝:剔除浮浆,并使其向下凹 10 mm,保证混凝土接缝处的质量,并充分湿润和冲洗干净,且不得积水。 4. 顶板施工缝:施工缝处顶板下铁垫 15 mm 厚木条,保证下铁钢筋保护层。上、下铁之间用木板保证净距,与上、下铁接触的木板侧面按钢筋间距锯成豁口,卡在钢筋上。 5. 施工缝处的混凝土浇筑:在施工缝处继续浇筑混凝土时,已浇筑的混凝土的抗压强度不应小于 1.2 N/mm^2。在浇筑混凝土前,宜在施工缝处铺一层与混凝土配比相同的水泥砂浆,接浆厚度 5 cm—10 cm。混凝土应仔细捣实,使新旧混凝土紧密结合。

2.1.4　防水工程质量控制措施

序号	质量控制点/项目	质量控制措施
1	卷材铺贴	1.基层处理剂涂刷均匀,涂刷完后达到干燥程度方可进行卷材铺贴。 2.屋面泛水卷材铺设时现行试铺,将立面卷材留足。先铺增强附加层,再铺平面卷材至转角处,然后从上往下铺贴立面卷材。
2	施工缝、后浇温度伸缩缝防水处理	增铺一层附加卷材
3	转角处防水处理	1.增铺与防水层同品种的卷材附加层。 2.在地下室外墙与平面交接处阳角增铺一层附加卷材(空铺)。

第四节　翻译评析

与文学翻译不同,国际官方招标文件逻辑缜密,内容准确,语言表达合乎逻辑。由于招标文件的叙述方式和工程英语的语言特点,翻译人员在翻译时要特别注意保持词汇、句法、篇章和文体上的一致性。

一、词汇层面

国际招标文件是一种法律合同,需要严格、规范运用词汇。合同双方经条款认可后,在保证期内应遵守所有条款。因此,招标文件内容的表述应准确,术语应专业。招标文件为技术标书,对准确性的要求高于其他文档。由于招标文件是由专业技术人员编写的,他们倾向于使用客观、直接、简洁的词汇和语言来描述项目,因此翻译应尽量保留词汇方面的特点。

1.频繁使用正式词语和短语

众所周知,招标文件的目的是证明投标人的能力和赢得竞标的信心,因此措辞应有利于塑造投标人认真细致的印象。此外,招标文件的内容主要是关于技术部分,因此,展示投标人的专业水平更加重要。

工程翻译中专业术语的翻译一般应统一,为了严谨起见,有时也会统一非专业词语,其中包括但不限于文件中出现的需统一的文件名、人名、地名、职称名、组织机构名等。尤其是在多人合译同一文件过程中,统一术语是非常重要的要求。在不同的国际文件中,招标和投标可使用"tender"和"bid"来表达,并无统一标准。在案例一的译文中,译者参考《中华人民共和国招投标法》双语版本,统一将招标译为"bid invitation",投标译为"bid submission",招标人译为"bid inviter",投标人译为"bidder"。

【例1】原文:

施工总工期:30日历天。

译文:Total construction period:30 calendar days。

分析:日历天(calendar days),是指日历上的日期,每一个日期算一天,意思就是一周按7天算。与之相对应的工作日,就是在计算时把一周时间算成5天。工作日一般是指除去法定节假日的时间,日历天是不除去法定节假日的自然天数。一般合同工期都是以日历天计算的。

在工程投标上有日历天和天之分,其含义是不同的。日历天就是过一天算一天,如工期100日历天就是100天。"天"在投标中的含义存在争议,因此目前大部分投标文件中的工期要求按日历天计算。

【例2】原文:13.工程发包价;14.合同承包价。

译文:13. Project Contract Issuing Price; 14. Contract Price。

分析:发包是指建设工程合同的订立过程中,发包人将建设工程的勘察、设计、施工一并交给一个工程总承包单位完成,或者将建设工程勘察、设计、施工的一项或几项交给一个承包单位完成的行为。建设工程的发包与承包,是从不同主体的视角对同一法律行为的描述。发包与承包是建设工程合同订立不可或缺的过程。

【例3】原文:Engineers will be randomly dispatched to check precision of the fabricated steel reinforcement bars at the suppliers workshop as per the drawings to ensure the dimensions are accurate.

译文:派遣工程师不定时地到供应商处按照图纸说明检查钢筋制作的精度,确保钢筋配料尺寸准确。

分析:第一句中的dispatch是一个正式单词,意思是"派某人去某地",尤其是出于某种特殊原因。选择用"派遣"是为了表明投标人已经准备派遣工程师对钢筋进行定期检查。

2. 常用词语的专业化

常用词语的专业化是指英语专业术语中的半技术性词语和部分一般性词语基本来源于英语普通词汇。这些常用词在一个特定的科学技术领域被引用。虽然它们被赋予了新的含义，但它们仍然与基本含义密切相关。为了熟悉专业术语，译者必须阅读大量相关领域的技术文本，这样才能使翻译更加专业。

【例4】原文：After the concrete is poured, the frame will be disassembled for circulating use before the vertical steel reinforcement bars are bound with the upper layer.

译文：混凝土浇筑完毕，上层柱竖向钢筋绑扎前将定距框拆卸，供周转使用。

分析：在这个句子中，"circulating"一词的原意是液体、气体或空气在一个地方或系统周围连续流动，在译文中却被赋予了"周转"的新意义。由于在接下来的施工过程中，定距框将在其他地方重复使用，因此"周转"一词可以表示定距框将被"取出"的意思。我们知道框架是平行四边形，不能简单地从混凝土中取出。而通过"拆卸"（"disassemble"），框架可以继续装配和循环使用。选择这两个词可以使表达更专业。

以单词"work"为例，这个词在翻译中有多种含义。比如：在物理学中，"work"的意思是"功"；冶金学家用"work"这个词来描述与冶金有关的各种复杂的过程，意思是"工序"。本文中"work"被翻译为"工程"。例如，"油漆工程""外部工程"和"金属工程"分别翻译为 paint work、external work 和 metal work。与"项目"这个词相比，"工程"更为具体。

另一个例子是单词"bar"。这个单词在词典里意义众多，最常见的意思是"酒吧""栏杆"。而根据本文中的工程项目投标文件的文本类型判断，"bar"在土木工程中的专业释义应为"钢筋"。由于钢筋的形状、性能和功能不同，"bar"与其他单词组合在一起，便形成了一些专业术语，如"slab bar""beam bar""stair bar"等。在上面的例子中，一些常见的单词被转换成罕见的术语，准确地表达了原文所传达的信息。

3. 大写中文数字、专有名词、项目编号的翻译

【例5】原文：闽建发招字〔20××〕×××号。

译文:Minjianfa Zhaozi〔20××〕×××。

【例6】原文:×政综〔20××〕×××号文。

译文:X.Z.Z Document〔20××〕No. ×××。

分析:项目、文件编号通常不必逐字翻译。英文字母构成的编号可以不翻,中文字构成的编号可以采用拼音首字母缩写或拼音全拼的方式译出。例5和例6分别采用拼音全拼和拼音首字母缩写的方式翻译编号。

【例7】原文:

投标保证金金额为人民币:壹万元整(￥10000元),投标人必须按规定交纳。

银行账户名称:×××。

开户行:建设银行×××支行。

账　号:×××××××××××××××××××。

译文:The bid security shall be RMB:壹万(Chinese numeral capital case) Yuan(￥10000 Yuan), which shall be paid by the bidder in accordance with the regulations.

Bank account name: ×××.

Opening bank: CCB ××× Branch Bank.

Account number: ××××××××××××××××××××.

分析:大写中文数字是中国特有的数字书写方式。利用与数字同音的汉字取代数字,多出现于银行的收据中,这样可以有效地防止数字被涂画。若将其翻译为英文数字则失去其特有功能,因此大写中文数字一般无须翻译。如第一次遇到可以增加注释进行说明。

遇到熟悉的专有名词,可用缩写词进行翻译,以达到简洁的目的。中国建设银行是人们熟知的中国国有银行之一,其缩写为CCB(Chinese Construction Bank)。

4.同义取其同

【例8】原文:本招标工程项目不允许违法转包、分包。除地基基础和主体工程外确需分包的,应符合有关法律、法规规定并经招标人同意。

译文:Illegal subcontracting are not allowed in this bidding project. If subcontracting is necessary except for foundation and main works, it shall comply with relevant laws and regulations and be approved by the bid inviter.

分析："转包""分包"意思相近,且在英语中对应同一个单词"subcontract",可采用同义取其同的翻译方法,翻译为"subcontract"。

二、句法层面

为了体现客观准确性,避免叙事性和主观描述性,投标文件中祈使句、陈述句、被动语态和各种复合句使用较多。在翻译时应注意其与中文句子在语法和语用方面的区别,达到工程文件句子饱满有力的目的。

1. 句式选择

英语句子的表达功能各不相同,因此在招投标文件的翻译中出现了句式选择的问题。由于招投标文件具有较强的表现力,叙述内容必须严谨、准确,声明应明确界定。为了使招投标双方准确地理解权利和义务,招投标文件更多地使用陈述句进行解释、规定和判断,使语言流畅、客观。因此,翻译时同样主要使用陈述句,不使用感叹句和疑问句,避免修辞格、双关语等修辞手段。需要注意的是,本文件为技术投标文件,陈述句特别是祈使句更适合说明技术要点。

以陈述句和祈使句形式写成的句子是客观的、权威的。文件中的绝大多数句子都是这种类型,目的是为了清楚地显示技术构建程序,避免技术过程中出现任何歧义。

【例9】原文:投标人应承担其编制投标文件与递交投标文件所涉及的一切费用,不管投标结果如何,招标人和招标代理机构对上述费用不负任何责任。

译文:The bidder <u>shall</u> bear all expenses involved in the preparation and submission of bid submission documents, regardless of the result of the bid, the bid inviter and the bidding agency <u>shall not</u> be liable for the above expenses.

分析:shall 在条约、规定、法令等文件中表示义务或规定。

【例10】原文:投标人对工程情况和招标文件(含工程量清单及工程预算价)有疑问的可于××××年××月××日 17:00 前直接登录××市招投标中心网向招标人提出(应列明招标项目名称和编号)。

译文:If the bidder has any questions about the project situation and the bid invitation documents (including the bill of quantities and the project budget price), it may submit to the bid inviter through ×× City Bidding Center website before 17:00

on date ×××（the name and number of the bidding project should be specified）。

分析：描述概念、事实和结果时，应当使用第三人称，第三人称强调的是所呈现的信息。通常情况下，工程技术文件的内容是与事情有关的信息，而不是与人有关的信息。因此，第三人称通常使用名词和第三人称代词表示，例如 it、they。使用第三人称时，不能使用指定性别的代词，例如 he 和 she，除非信息表述中要求这样使用。

【例11】原文：

3. 投标人的资格资质要求

3.1 投标人资质等级要求具备建设行政主管部门核发的水利水电工程施工总承包三级及以上（不分主、增项差别），并具有有效的安全生产许可证的企业。

3.2 凡在××市内注册登记，具有独立法人资格，符合规定资质条件且未因违法违规被限制参加相应项目投标的潜在投标人均可参与投标。

3.3 近三年内，未被本市县区级以上相关行政监督部门记入不良行为的。

译文：

3. Qualification Requirements of Bidders

3.1 The bidder's qualification level requires the enterprise with Water Conservancy and Hydropower Engineering Construction General Contracting Level 3 or above issued by the construction administrative department（regardless of the difference between the main and additional items）and valid safety production license.

3.2 Any potential bidder who is registered in ×× City, has an independent legal personality, meets the prescribed qualification conditions and is not restricted from participating in the corresponding project bidding due to violations of laws and regulations may participate in the bidding.

3.3 In the past three years, the bidder has not been recorded as bad behavior by the relevant administrative supervision departments at or above the county and district level of the city.

分析：本部分为投标人资格资质描述，句子较长，信息较为密集，但是句子结构并不复杂，由简单的平行句组成。翻译时采用顺序法，根据原文句子顺序翻译即可。要特别注意的是，对于资格资质的表达，绝不可错译、漏译，否则对投标人可能造成巨大的影响。

2. 多用复合句

招投标文件信息量比较大,因此翻译时多用复合句,因为复合句既可以明确定义,又可以避免不必要的重复。有一些复合句包含多种类型的复合成分或多个分句,以明确双方的权利和义务,进一步解释或限制互补,避免曲解和误解。

【例 12】原文:After the steel bars are transported to construction site, the steel fixers will be in charge of the management to ensure that the steel reinforcement bars are sorted, kept and used in accordance with the applied parts and time.

译文:钢筋加工完毕运至现场后设专人负责成型钢筋的管理工作,保证成型钢筋按使用部位、使用时间进行分类分批编组编号码放、保管及发放。

分析:例 12 包含一个宾语从句,用于清楚地说明将钢筋运送到站点后的步骤。在翻译这类汉语句子时,译者需要灵活运用不同的技巧,做到语义等价。

【例 13】原文:投标人应按招标文件规定的金额,将投标保证金从投标人银行基本账户按招标项目单标单笔一次性汇入招标文件指定的银行账户。若因投标人分批分笔缴交投标保证金而导致的投标保证金统计失误的,由投标人自行负责。

译文:The bidder shall, according to the amount specified in the bid invitation documents, transfer the bid security from the basic account of the bidder's bank to the designated bank account in the bid invitation documents in one lump sum according to the single bid project. The bidder shall be responsible for any statistical error of bid security caused by the payment in batches.

分析:中译英遇到复杂句时同样需要进行结构分析,以正确理解原文的含义。本句中主句为"投标人将投标保证金汇入银行账户"。其中:"按招标文件规定的金额"为状语,说明汇款的金额要求;"从投标人银行基本账户"为状语,说明保证金来源;"按招标项目""单标单笔""一次性"均为状语,说明汇款的方式;"招标文件指定的"修饰限定"银行账户"。

通过分析可知,本句状语较多、较复杂,翻译时需要拆分、变换位置,以使译文通顺流畅。因此,把"according to the amount specified in the bid invitation documents"作为状语插入主句中,与主语靠近,增强它与主语的逻辑关系;把"从投标人银行基本账户"翻译为地点状语"from the basic account of the bidder's bank"，与主句宾语靠近，说明其与宾语的直接修饰关系；把"单标单笔""一次性"翻译为句尾连续方式状语"in one lump sum according to the single bid project"，以修饰整个句子。

本句重复提及的"投标保证金"和"投标人"在翻译时可省略。

三、篇章层面

众所周知，篇章是表达整体概念的语义单位。它不仅是句子和单词的集合，还是一个结构完整、功能明晰的语义统一体。在翻译过程时，应将篇章整合起来，使译文与原文在整体上保持一致，而不应只注重词语的表面意义和等值，而应努力实现原文与译文在篇章层面的对等。

由于英汉语言之间的差异，译者不能孤立地翻译原文中的单词和句子，而要在语境中分析原文的内容。

【例14】原文：Before pouring the concrete, it is advisable to pave a 5 – 10 cm layer of cement mortar with the same mixing proportion at the construction joints. The concrete shall be compacted to be fully mixed.

译文：在浇筑混凝土前，宜在施工缝处铺一层与混凝土配比相同的水泥砂浆，接浆厚度5 cm～10 cm。混凝土应仔细捣实，使新旧混凝土紧密结合。

分析：在原文中，"the concrete"指意不明，内涵模糊。但是前文中明确表示，需铺一层相同配比的水泥砂浆后再浇筑混凝土。由此可知，"the concrete"是指紧密结合的新旧两层混凝土。

【例15】原文：The enhanced additional layer shall be paved first, then the plane coil to the corner and finally the facade roll from upper to bottom.

译文：先铺增强附加层，再铺平面卷材至转角处，然后从上往下铺贴立面卷材。

分析：原文中，"pave""coil"和"roll"三个动词意义不同。尽管"coil"和"roll"表意生动，但是在汉语中很难找到等值的词语进行翻译。因此，在译文中，译者重复使用动词"铺"，与宾语"增强附加层""平面卷材"和"立面卷材"有较好的搭配适应性。

第五节　拓展延伸

一、中译英

（一）投标文件的编写

1. 投标文件的组成

1.1　投标文件须按招标文件提供的统一格式填写（格式中的所有表格，均可视内容进行扩展），包括完整地填写投标函，中文打字并装订成册。

1.2　投标文件由下列内容组成：

（1）投标承诺函；

（2）建造师或临时建造师证书复印件；

（3）相关证件原件。

（二）投标文件的递交

2. 投标文件的密封及标记

投标文件（正本一份、副本三份）应由两个内包封组成：一是投标文件内包封；二是招标文件要求提供的证件原件内包封。所有内包封均应密封完好并在封口处加盖投标人公章，再将两个内包封一起放入一个外包封中。外包封上应写明：①招标人名称；②招标编号；③"开标时间前不得开封"的字样；④投标人的名称、地址与邮政编码。投标文件外包封应密封完好并在接缝处加盖投标人公章，并由投标人法定代表人或其授权代理人盖章，否则招标人不予接收。投标人在递交投标文件时另提交一份投标保证金缴交凭证复印件和银行开户许可证复印件（加盖投标人公章），用于退还投标保证金。

3. 投标文件的提交和投标截止时间

3.1　投标文件提交截止时间为×××年××月××日15:00时。提交投标文件的地点为×××。

3.2　投标人须随带个人身份证、法定代表人资格证明书或法定代表人授权委托书、投标保证金缴交凭证复印件（已按×招投标办〔×××〕××号文缴交投标保证金的不须提供）参加开标会。

3.3　存在下列情形之一的，招标人不予开启其投标文件并予以退还：

（1）投标文件逾期送达的；

（2）投标文件外包封未按要求密封或盖章的；

（3）投标人代表未随带有效身份证的；

（4）投标人未提交法定代表人资格证明书或法定代表人授权委托书的；

（5）法定代表人资格证明书或法定代表人授权委托书未按要求签字、盖章的；

（6）未按规定交纳投标保证金或投标保证金缴交凭证和指定银行对账单上均没有体现投标人、投标项目编号、投标项目名称的。（已按×招投标办〔××××〕××号文缴交投标保证金的不须提供）

（三）开标、评标、定标

4. 开标

4.1　开标会定于××××年××月××日 15:00 时在××进行，开标会由招标人或招标代理机构主持并邀请行政监督部门参加，投标人的法定代表人或其委托的代理人必须参加开标会。

4.2　开标程序：

（1）招标人或招标代理机构介绍参加开标会的有关单位；

（2）确定进入公开抽取程序的合格投标人名单，并由招标人当场公布；

（3）随机抽取确定中标人。

5. 评标

5.1　评委对投标人的资格、资质和投标文件的有效性进行审查。

5.2　投标文件或投标人存在下列情形之一的，应当作为不合格标处理，不得进入随机抽取程序：

5.2.1　投标人没有在××市内注册登记、没有独立法人资格的承包商企业；

5.2.2　投标人的企业资质不符合招标文件要求；

5.2.3　投标承诺函未按要求格式填写或签字、盖章的；

5.2.4　投标人的承诺不满足招标文件要求的；

5.2.5　未提交建造师或临时建造师证书复印件或未加盖投标人公章的；

5.2.6　近三年内，曾被县区级以上相关行政监督部门记入不良行为的；

5.2.7　未按招标文件要求提供相关五种证件原件（营业执照副本、资质证书副本、建筑施工企业安全生产许可证副本、银行开户许可证或开户核准通知书、交纳投标保证金凭证)(已按×招投标办〔××××〕××号文缴交投标保证金的不须提供)进行核对的。

6.确定中标人

6.1 从审查合格的投标人中随机抽取一家中标人。

6.1.1 招标人或招标代理机构宣布抽取规则;

6.1.2 按提交投标文件的顺序,由每个审查合格的投标人公开抽取、登记代表该投标人的号码,并当众公布每个号码所对应的投标人;

6.1.3 招标人公开随机抽取其中一个号码,该号码所对应的投标人即为中标人;

6.2 招标人或招标代理机构撰写开标情况报告,由招标人、招标代理机构及相关行政监督部门签字确认后入档。

6.3 经过审查,只有一家合格投标人的,招标人可以直接指定该投标人为中标人。

6.4 招标人应当自确定中标人之日起,将招标项目名称、中标人名称及中标金额在××市招投标中心网站和招标人所在单位公示。

7.中标通知

在开标完成后2日内根据中标结果,招标人以"中标通知书"的形式通知中标人。

(四)授予合同

8.签订合同

8.1 招标人与中标人应于中标通知书发出之日起5日内,按照中标通知书、招标文件和中标人的投标文件签订建设工程施工合同。双方不得再进行订立背离合同实质性内容的其他协议。

8.2 招标人应在签订施工合同后15日内将施工合同原件提交相关行政监督部门备案。

8.3 中标单位逾期不与建设单位签订施工合同的,取消中标单位的中标资格,投标保证金不予退还。

9.履约担保

中标人在签订施工合同时,应向招标人提交履约保证金,履约保证金数额为合同金额的10%。履约保证金必须用现金汇入招标人的指定账户。

二、英译中

Chapter Four   Health and Safety Plan

4. 1 Safety Construction Objectives

Avoid severe human injury and machinery accidents, and control the frequency of the industrial injuries below 3%.

4. 2 Safety Guarantee Guidelines and Principles

(1) Safety Guarantee Guidelines:

During the execution of the contract, our company will observe the relevant safety standards applicable to both Chinese and Botswana government, abide by the policy of "safety first and precaution utmost", and accept the the field supervision engineer's instructions and supervision.

(2) Safety Production Principles:

①In case of contradiction between production and safety, adhere to safety first;

②Insist on the principle that production and safety must be managed simultaneously;

③Adhere to the principle of synchronous implementation of production and safety.

4. 3 Construction Safety Guarantee System

4. 3. 1 Establish and Perfect the Safety Management and Organization Unit

After signing the contract, we will immediately select the experienced × × × officer, set up the project safety management team and strictly enforce the safety construction procedures to smoothly and safely fulfill the project contract.

4. 3. 2 Formulate the Construction Safety System and Regulation

(1) Construction safety responsibilities and agreements:

All field working staff must understand their safety responsibilities, and the × × × officer and subcontractor sign the safety agreement; Staff without the safety agreement are not allowed to work in the project; The safety agreement is used to standardize the construction safety work.

(2) Safety Training System:

All working staff shall receive special safety training and tool-box training on a regular basis. The safety training records shall be submitted to the project manager for inspection.

(3) Safety Technology Education System:

Before starting all parts of work, the technical director shall introduce general technology and construction methods and safety technical measures to the construction staff to make them understand the safety matters in the work.

(4) Safety Check System:

The × × × officer will irregularly check the safety conditions at the construction site and timely prepare and submit the safety report and formulate the safety measures.

(5) Safety Meeting System:

Safety meetings are held periodically during the execution of the contract. The construction safety report is amended and discussed at the meetings, and relevant safety measures are adopted to ensure the smooth operation of the construction.

(6) Safety Report System:

The safety report and meeting minutes shall be submitted to the project manager and the site manager for periodic check. In case of any accidents, relevant measures shall immediately apply.

4.4 Guarantee Measures for Construction Safety Technology

4.4.1 Safety Protection for the Construction Staff

(1) Construction staff who enter the construction site must receive the safety training and pass the examination.

(2) The construction staff must observe the field discipline and the national codes, laws and regulations, and comply to the comprehensive management of the project department.

(3) The construction staff must wear Safety Gear (PPE) at all times when entering the construction site; and must be attached to safety belts when working on frames higher than 2 m.

(4) The construction staff who work at high altitude mustn't bare foot or back, and mustn't wear slippers or hard soled shoes.

(5) The construction staff are not allowed to dismantle all safety protection facilities on the site, such as mechanical protection shell, safety net, safety fence, external frame joint, warning signal and etc. If required, the × × × officer and the field supervisor's consent shall be required.

(6) Two dysprosium lamps are installed on the tower body for night construction and iodine-tungsten lamps are installed in allocated areas. Sufficient electric lights are installed in all work areas to assure safety of the construction and construction staff.

(7) In the process of the project construction, periodically spray the permitted pesticide for killing insects and mice to prevent injury of other staff.

4.4.2 Safety Protection Measures for Scaffolding:

(1) The scaffold foundation must be firm and satisfy the load requirements. Herringbone strutting and protection guardrail are arranged in accordance with the construction specifications, and drainage and anti-slippery measures at the scaffold location must be well done on rainy days.

(2) The scaffold erection height should be 1.5 m higher than the operation surface at the top of the building, and protection guardrail is provided.

(3) Steel tube, fastener, scaffold board, etc. on the scaffold mustn't be dismantled. In case of any special circumstances when it has to be dismantled, corresponding remedial measures must be specified and implemented by the engineer and the × × × officer. After the process is completed, the dismantled steel pipe, fastener and scaffold should be restored immediately.

(4) The scaffold needs checking and it can't be used until the qualification is confirmed and the acceptance sheet is filled in. The sedimentation of the scaffold shall be observed after the rain and reinforcement measures shall be adopted in case of abnormality.

(5) Stairs, slopes, inclined slopes and climbing ladders of the upper and lower scaffolding must be provided with handrails or other safety protection measures and barriers in the channel must be cleared to assure staff safety.

(6) Establishing and dismantling the scaffolding shall be carried out after the

disassembling plan is prepared and approved.

(7) Nothing shall be piled up on the scaffold. The scaffold must be firmly connected with the main body of the building to prevent the overall instability.

(8) When the scaffolding is dismantled, the professional construction engineer should first check the connections with the structure, and transfer the retained material and sundries on the scaffold to the ground by manual. The dismantling process should be carried out from top to bottom and in the order of disassemble. Throwing objects downwards from the top must be prohibited during the dismantling operation and high altitude construction.

(9) The scaffold must be reliable and firm with armrest on the upper side. The width and thickness of the plank must ensure the safety of construction, and the probe plate must be fastened.

第四章　工程合同文本

工程合同指工程建设中的各平等主体之间，为达到一定的目标而明确各自权利义务关系的协议。工程合同在工程建设中扮演着重要的角色，良好的工程合同不仅是工程项目建设质量的保障，更是促进工程项目市场良性运转的基础。

第一节　背景分析

随着中国经济的快速发展，以及经济的全球化，中外企业工程合作项目数量也达到了历史新高。国际工程项目承包是指一个国家的政府部门、公司、企业或项目所有人（一般称工程业主或发包人）委托国外的工程承包人负责按规定的条件承担完成某项工程任务。国际工程承包是一种综合性的国际经济合作方式，是国际技术贸易的一种方式，也是国际劳务合作的一种方式。之所以将国际工程承包作为国际技术贸易的一种方式，是因为国际承包工程项目建设过程中，包含有大量的技术转让内容。特别是项目建设后期，承包人要培训业主的技术人员，提供所需的技术知识（专利技术、专有技术），以保证项目的正常运行。

在国家提出"一带一路"倡议后，中国企业纷纷"走出去"，国际工程项目越来越多。比如，中国企业依托国内大规模铁路项目建设经验，以及国家提出的"一带一路"和中国铁路"走出去"的倡议的强有力政策支撑，在国际铁路工程咨询、建设管理、装备制造领域获得了一定的市场，截至 2019 年，已在世界 50 多个多家和地区参与铁路建设项目。

国际工程是我国对外经济合作的重要内容。工程项目建设周期长、涉及因素多、专业技术强，当事人之间的权利、义务关系十分复杂，不是靠简单的口头约定就能解决问题。近年来，国际工程市场也发生了很大变化，合作条件越来越严格,法律和合规风险越来越大,如围标串标风险、贿赂与腐败风险、出口管制风险、环境生态风险、合同责任风险、融资风险、担保风险、劳动用工风险、税务风险、外汇监管风险,以及各种疫情和经济危机对国际工程项目造成的拖期与成本控制风险等,这些风险都需要我国企业从合同、法律和合规的途径进行预防和控制。为了最大限度地监管施工过程、保护双方利益,工程项目的实施需要通过各种合同来进行规范和运作。作为在双方自愿的基础上明确权利与义务的正式法律文件,工程合同在国际工程项目中的地位与日俱增。

一、工程合同分类

工程合同的形式和类别非常之多。工程合同的分类方法有很多,如:按工作内容可分为工程咨询服务合同(包括设计合同、监理合同等)、勘察合同、工程施工合同、货物采购合同(包括各类机械设备采购、材料采购等)、安装合同、装修合同等;按承包范围可分为设计建造合同、EPC/交钥匙合同、施工总承包合同、分包合同、劳务合同、项目管理承包合同等。

二、合同翻译原则

1. 忠实准确原则

在拟定合同时对合同中的每一项条款、每一个词语都必须最大限度地保证精准,因为合同中的一个词语,都可能成为日后产生纠纷时双方进行争论的重点,因此在合同翻译中必须遵循准确原则。

忠实准确原则首先体现在对词义和内容的准确把握上,要求译者在透彻理解原文的基础上运用恰当的词语和结构将原文转换为译文。任何歪曲和疏漏都有可能成为合同任何一方当事人逃避法律责任的借口,因此任意增减内容、有意或无意地误译在工程合同的翻译中都是不允许的。

忠实准确原则不仅适用于词义和内容,也适用于文体风格。工程合同经常使用词汇、句法等手段体现其作为法律文书的正式、庄重等特点,因此译者在翻译过程中应当选择相应风格的词句、结构等,使译文的风格与原文的风格相符合,做到在风格上忠实于原文。

2. 规范通顺原则

所谓规范通顺,就是用规范通顺的、合乎合同语言要求的文字表达出来,因为合同属于严肃的文体,不允许文字上的随意性。合同通常是由具有商务和法律方面专门知识的人起草的、适应当事人需要的一种法律文书。因此,英语商务合同具有商务和法律方面的专业特征,对词汇、句法都有特定的要求和规范,如使用专业词语、特定的句式等。所以,在翻译工程合同时,译者必须遵循规范原则再现原文的特征,使译文符合工程合同的规范。不仅如此,翻译工程合同时还要做到语言通顺,反复探索各合同条款间的逻辑关系,确保语句结构合理。

第二节 专业术语

序号	中文	英文
1	包括但不限于	including but not limited to
2	包装与运输	packaging and transportation
3	保留金	retention money
4	保密事项	confidential details
5	保密性	confidentiality
6	保险	insurance
7	保险凭证	insurance certificate
8	保障	indemnities
9	保证条款	warranty clause
10	变更	variation
11	变更程序	variation procedure
12	变更令	variation instruction
13	不可抗力	force majeure
14	不可预见的客观条件	unforeseeable physical conditions
15	财务报表	financial statements
16	财务结算	financial settlement
17	裁决规则	rules for adjudication
18	采购	procurement
19	差额	balance

续表

序号	中文	英文
20	成本指数	cost index
21	承包商	contractor
22	承包商代表	contractor's representative
23	承包商设备	contractor's equipment
24	承包责任制	contract responsibility system
25	承兑交单	documents against acceptance
26	承兑期	acceptance period
27	承租人	lessee
28	出租人	lessor
29	单位工程	section
30	对价	consideration
31	发货期	delivery time
32	法定权益	legitimate rights and interests
33	法定义务	binding obligations
34	放线	setting out
35	分包商	subcontractor
36	付款币种	currencies for payment
37	付款交单	documents against payment
38	付款证书	payment certificate
39	附加条款	additional clause
40	复工	resumption
41	工程	works
42	工程合同	engineering contract
43	工程接收	taking over
44	工程量表	bill of quantities
45	工程师	engineer
46	工程造价	project cost
47	工期延长	extension of time

续表

序号	中文	英文
48	工伤保险	job injury insurance
49	工业产权	industrial property rights
50	工作日	workday
51	估价	evaluation
52	雇主/发包商	employer
53	雇主代表	employer's representative
54	雇主设备	employer's equipment
55	关联公司	affiliate
56	国际咨询工程师联合会	FIDIC
57	合同标的	contract object
58	合同的各方	party/parties
59	合同法	contract law
60	合同副本	copies of the contract
61	合同价格	contract price
62	合同金额	contract amount
63	合同期满	expiration of contract
64	合同期限	duration of contract
65	合同效力	validity of contract
66	合同协议书	contract agreement
67	合同续订	renewal of contract
68	合同正本	originals of contract
69	合同终止	termination of contract
70	合同转让	assignment of contract
71	和解	conciliation
72	货到付款	pay on delivery
73	货物	goods
74	基准日期	base date
75	计日工作	daywork

续表

序号	中文	英文
76	甲方	Party A
77	价格细目表	price breakdown
78	价值工程	value engineering
79	检验	inspection
80	建筑许可证	construction permits
81	交货地点	delivery place
82	接收证书	taking-over certificate
83	结清证明	discharge
84	进度报告	progress reports
85	进度计划	programme
86	进度款	progress payment
87	经济责任	financial responsibility
88	举证责任	burden of proof
89	拒付	dishonor
90	拒收	rejection
91	竣工报表	statement at completion
92	竣工后试验	tests after completion
93	竣工时间	time for completion
94	竣工试验	tests on completion
95	开工	commencement
96	开工日期	commencement date
97	宽限期	grace period
98	劳动合同期限	term of a labor contract
99	连带责任	joint liability
100	临时工程	temporary works
102	履约担保	performance security
103	履约能力	contractual capacity
104	履约证书	performance certificate

续表

序号	中文	英文
105	民事诉讼	civil litigation
106	目的口岸	port of destination
107	期满日期	expiry date
108	期中付款证书	interim payment certificates
109	其他条款	miscellaneous clause
110	签发	sign and issue
111	侵权	infringement
112	清算人	liquidator
113	权利	entitlement
114	权益转让	assignment
115	缺陷通知期限	defects notification period
116	缺陷责任	defects liability
117	日工作计划表	daywork schedule
118	日历天	calendar day
119	融资费用	financing cost
120	社会保险和福利	social insurance and welfare
121	审计报告	audit report
122	生产设备	plant
123	生效	come into effect
124	实物支付通知	payment in kind notice
125	适用法律及执行	governing law and enforcement
126	受让人	assignee
127	受托人员	delegated persons
128	受益人	beneficiary
129	授权签名	authorized signature
130	书面函件往来	communications in writing
131	税务抵减	deduction of tax
132	索赔	claims

续表

序号	中文	英文
133	特殊条款	special provisions
134	调价公式	price fluctuation formula
135	调解	mediation
136	投标函	letter of tender
137	投标人	tenderer
138	投标书附录	appendix to tender
139	土地使用费	royalties
140	拖长的暂停	prolonged suspension
141	外币	foreign currency
142	违法行为	illegal act
143	违约	default
144	违约赔偿金	liquidated damages
145	违约条款	default clause
146	未支付的/未完成的	outstanding
147	文件的照管和提供	care and supply of document
148	文件优先次序	priority of document
149	无效	null and void
150	现场	site
151	现场数据	site data
152	现场进入权	right of access to the site
153	现场清理	clearance of site
154	项目经理	project manager
155	销售确认书	sales confirmation
156	信用证	letter of credit
157	刑事诉讼	criminal litigation
158	修补工作	remedial work
159	验收条款	inspection clause
160	要约	offer

续表

序号	中文	英文
161	一般条款	general provisions
162	乙方	Party B
163	隐蔽工程	concealed work
164	佣金	commission
165	永久工程	permanent works
166	优惠条款	preferential clause
167	友好解决	amicable settlement
168	有缺陷的工程	defective works
169	有效	validity
170	预付款	advance payment
171	暂列金额	provisional sum
172	暂时停工	suspension of work
173	责任	liability
174	责任限度	limitation of liability
175	债权人	creditor
176	债务人	debtor
177	招标人	tenderee
178	争端	disputes
179	争端裁决委员会	dispute adjudication board
180	整改	rectification
181	支付条款	terms of payment
182	知识产权	intellectual property rights
183	执照	licences
184	指定的分包商	nominated subcontractor
185	制裁	sanction
186	质量保证文件	quality assurance documents
187	质量监督	quality supervision
188	中标函	letter of acceptance

续表

序号	中文	英文
189	中止履行合同	suspension of the contract performance
190	终止	termination
191	仲裁	arbitration
192	重置成本	replacement cost
193	转嫁的过失责任	imputed negligence
194	转让人	assignor
195	装运口岸	port of loading
196	装运通知	shipping notice
197	子公司	subsidiary
198	最终报表	final statement

第三节　翻译案例

原文：

Contract Documents

The following documents shall constitute the contract between the employer and the contractor and each shall be construed as an integral part of the contract. During the period of contract execution, written agreement(s) or document(s) in relation to negotiation/consultation, modification of works between employer and contractor shall be deemed as parts of contract.

1. Intent of the Contract

The contract shall not be modified or altered unless otherwise agreed in writing by both parties. The contract intents to cover and provide for first class completed work in all respects and everything necessary to carry out this intent, which may be reasonably implied from the contract, even if not particularly referred to in the contract.

2. Scope of Work

The scope of work herein includes but not limited to: coal handing system construction and subsidiary facilities foundation, coal yard foundation, warehouse and maintenance house, ESP control house and air compressor room. The employer shall have the right to adjust the scope of the works, but shall notify the contractor in writing, and the contract shall accept the project unconditionally.

3. Definition

The capitalized words and phrases used herein shall have the same meanings as described to them in the contract terms.

4. Payment

The employer shall pay the contractor in consideration of the performance of the works such sum as may become payable under the provisions of the contract.

5. Precedence

In the event of any ambiguity or conflict between the documents forming this contract, the order of precedence shall be the order in which the documents are listed in item 1 of this contract agreement.

6. Employer's Authority to Delegate

Employer may from time to time delegate to the engineer any of the duties and authorities vested in employer as specified in contract and may any time revoke such delegation. Any such delegation or revocation in writing shall, prior to 7 (seven) days, be given to contractor. Notification of such delegation or revocation shall be attached to this contract.

7. Employer's Instruction

Instructions given by the employer shall be in writing, provided that if for any reason the employer considers it necessary to give any such instruction orally, the contractor shall comply with such instruction. Confirmation in writing of such oral instruction given by the engineering, whether before or after the carrying out of the instruction, shall be deemed to be an instruction within the meaning of this subclause. Provided further that if the contractor, within 7 (seven) days, confirms in writing to the employment any oral instruction of the employer and such confirmation is not contradicted in writing within 14 (fourteen) days by the employer, it shall be deemed to be an instruction of the employer.

Contractor shall, if deems the employer's instructions unreasonable, present modification statement of such instructions in writing to the employer within 24(twenty-four) hours upon receipt of instructions. The employer shall, within 24 (twenty-four) hours upon receipt of such statement thereof, determine to modify or maintain the instructions in written form. In case of urgent circumstance, contractor shall execute instructions hereof the employer requiring contractor to execute forthwith or instructions the employer determining to maintain notwithstanding contractor dispute. Additional payment incurred out of incorrect instruction and damage hereof caused shall be borne by employer. Construction schedule thereof delayed shall be extended accordingly.

8. Employer's General Works

8.1 Perform the works of expropriation of land, removal of buildings thereon, compensation of removal therefrom, etc. ; make the land in the circumstance of be ready for construction and be responsible for handling the above said issues left over.

8.2 Provide contractor with data of engineering geology and underground pipelines.

8.3 Set benchmark and coordinates control point and send written form of which to contractor for acceptance on site; and be liable for assuring the trueness and accuracy of the data herein.

8.4 Organize drawings examination and design disclosure and presentation attended by contractor and design institute.

8.5 Delegate engineer representative to supervise project progress, quality and safety, to inspect hidden project, to conduct interim taking-over and project inspection, and to handle visa.

8.6 In the circumstance that project site is constructed by several companies, employer shall be entitled to organize coordination seminar which shall be attended by contractor(s) on time. Contractor(s) hereto shall be subject to employer's coordination and diligently execute the decision therefrom.

8.7 The employer shall make available to the contractor such access as may be required to enable the contractor to proceed with the execution of the works in accordance with the contract and may assist contractor in applying to the authorities concerned for such official permission.

9. Failure to give possession

If the contractor suffers delay and/or incurs costs from failure on the part of the employer to give possession in accordance with the terms, the engineer shall, after due consultation with the employer and the contractor, determine:

(a) any extension of time to which the contractor is entitled;

(b) the amount of such costs, which shall be added to the contract price.

10. Loss or Damage Due to Employer's Risks

In the event of any such loss or damage happening from any of the risks defined in sub-clause, or in combination with other risks, the contractor shall, if and to the extent required by the engineer, rectify the loss or damage and the engineer shall determine an addition to the contract price in accordance with clause. The engineer shall notify the above decision to the contractor, with a copy to the employer. In the case of a combination of risks causing loss or damage, any such determination shall take into account the proportional responsibility of the contractor and the employer.

11. Contractor's General Works

11. 1 The employer shall make the main contract(other than the details of the prices thereunder) available for inspection to the contractor. The contractor shall be deemed to have full knowledge of the provisions of the main contract. The contractor shall assume and perform hereunder all the obligations and the liabilities of the employer under the main contract in relation to the works.

11. 2 The contractor hereby acknowledges that any breach by him of the contract may result in the employer committing breaches of and becoming liable in damages under the main contract and other contracts made by him in connection with the works and may incur further loss or expense to the employer in connection with the works. In such cases, he shall indemnify the employer against any damages for which the employer becomes liable under the main contract as a result of such breaches.

译文：

合同文件

下列文件应构成适用于发包方与承包方间的合同,并各自形成完整合同的有关部分。合同履行期间,发包方与承包方间有关工程的洽商、变更等书面协议或文件视为本合同的组成部分。

1. 合同意向

若无合同当事双方书面同意,该合同不得进行修改或改动。本合同用于涵盖并提供为完成该意向所有方面及一切有关事项而进行的一切优质完整的施工,即便合同无特别阐述也应进行合理说明。

2. 工作范围

工作范围包括但不限于:输煤系统建设以及附属设备基础、煤场基础、材料库及检修楼、电除尘配电室与空压机房的全部建筑工程。发包方有权适当调整工程范围,但应书面通知承包方,承包方应无条件接受。

3. 定义

此处所用的大写词语及词组应与合同条款中描述的词语词组具有同等含义。

4. 付款

发包方应根据工程进度情况进行付款,在合同有关条款下,上述款项应为可支付款项。

5. 优先顺序

在形成合同的各种文件之间有语义含混或相互冲突的情况下,优先顺序应如合同协议书分项条款 1 中所列的文件顺序相一致。

6. 发包方的委派机构

发包方可随时委派工程师,行使合同约定的自己的职权,并可在认为有必要时撤回委派。委派和撤回均应提前 7(七)天以书面形式通知承包方。委派书和撤回通知作为本合同附件。

7. 发包商的指令

发包方应以书面形式发出指示,如果发包方认为由于某种原因有必要以口头形式发出任何此类指示,承包方应遵守该指示。工程师可在该指示执行之前或之后,用书面形式对其口头指示加以确认,在这种情况下应认为此类指示是符合本分条款规定的。如果承包方在7(七)天内以书面形式向发包方确认了发包方的任何口头指示,而发包方在14(十四)天内未以书面形式加以否认,则此项指示应视为是发包方的指示。

如果承包方认为发包方指令不合理,应在收到指令后24(二十四)小时内向发包方提出该指令的修改声明,发包方在收到承包方声明后24(二十四)小时内做出修改指令或继续执行原指令的决定,并以书面形式通知承包方。紧急情况下,发包方要求承包方立即执行的指令或承包方虽有异议,但发包方仍决定继续执行的指令,承包方应予以执行。因指令错误发生的追加合同价款和给承包方造成的损失由发包方承担,延误的工期相应顺延。

8. 发包方一般工作

8.1　办理土地征用、拆迁补偿等工作,使施工场地具备施工条件,在开工后继续负责解决以上事项遗留问题。

8.2　向承包方提供工程地质和地下管线资料。

8.3　确定水准点与坐标控制点,以书面形式交给承包方,进行现场交验,并有责任确保此数据的真实准确。

8.4　组织承包方和设计单位进行图纸会审和设计交底。

8.5　委派工程师代表,对工程进度、质量、安全进行监督,检查隐蔽工程,办理工程中间交工、工程验收及签证手续。

8.6　在现场有几个承包单位同时施工的情况下,发包方有权组织协调会,承包方应按时参加,服从发包方的协调并认真执行会议决定。

8.7　发包方应当给予承包方根据合同进入工程施工现场所需的入场便利措施,并协助承包方申请由权威机构颁发的这些正式许可证书。

9. 未能移交现场

如果由于业主未按条款规定让承包方移交现场,使承包方延误了工期和(或)招致费用,则工程师应在及时与发包方和承包方协商之后,决定:

(i)承包方有权获得延长工期;

(ii)应在合同价中增加此类费用总额。

10. 由于发包方风险造成的损失或损坏

当任何此类损失或损坏是由于分条款所限定的任何风险造成的,或是与其他风险相结合造成时,若工程师提出要求,则承包方应按该要求的程度修补这些损失或损坏。工程师应按照条款的规定,决定增加合同价的金额,并应相应地通知承包方,同时将一份副本呈交给发包方。如果是由多种风险相结合造成的损失或损坏,工程师在做出上述决定时,应考虑到承包方和发包方的责任所占的比例。

11. 承包方一般工作

11.1　发包方应将主合同交与承包方查阅(合同中提及价格细节部分除外)。承包方被认为就主合同有关规定已完全知晓。承包方应在主合同条款下,全面承担并履行与本工程相关的对发包方的义务及责任。

11.2　承包方应承认其任何违背合同的行为均可引起发包方违背主合同并对合同项下工程及其他由发包方签订的合同项下的有关工程的毁损负有责任,并可导致发包方产生与工程有关的进一步损失或开支。在此情况下,承包方应向发包方赔偿因其违背合同而引起的主合同项下发包方应负责承担的毁损。

第四节　翻译评析

本案例为典型的工程合同文本,提及了发包商和承包商的一般工作,以及因风险造成损失或损坏发生时各自应承担的责任。文本语言正式庄重,结构严谨,避免使用浮华或带感情色彩的词语,具有显著的公文文体特征。

一、词法层面

1.古体词的使用及翻译

国际工程合同经常涉及古体词。古体词较少有联想意义,语义比较明确,符合合同文体的严谨性要求,可以奠定全文严肃的行文基调,增强读者对合同文本的尊重程度。工程合同中用得最多的、最有特色的古体词大多是地点副词+类似介词的后缀组成的复合词,如 hereto、herein、hereby、therein、thereto、hereof、hereinbefore、thereof 等。这些词在现代生活中几乎已经被淘汰,但是在工程合同中时常出现。

【例1】原文:The capitalized words and phrases used herein shall have the same meanings as described to them in the contract terms.

译文：此处所用的大写词语及词组应与合同条款中描述的词语词组具有同等含义。

【例2】原文：Additional payment incurred out of incorrect instruction and damage hereof caused shall be borne by employer.

译文：因指令错误发生的追加合同价款和给承包方造成的损失由发包方承担。

【例3】原文：Contractor(s) hereto shall be subject to employer's coordination and diligently execute the decision therefrom.

译文：承包方对此应服从发包方的协调并认真执行会议决定。

类似的古体词还有：

hereinafter = later in the same contract 在下文

hereafter = after this 从此以后；今后

hereby = by means of/by reason of this 以此方式；特此

hereto = to this 本文件的

hereunder = under the clause/contract 在本条款/协议之下

hereof = of this 就此；以此

herein = in this 此中；于此

thereby = by that means 因此；从而

thereafter = after that, afterwards 此后

thereunder = under that part of a contract 在其下；按规定条款

thereof = of that 在其中；由此

whereby = by which 凭此协议/条款

2. 情态动词的使用及翻译

在工程合同中，为了明确双方的责任和权利，一般会出现"禁止、不得"或"可以、应当"等意思的表达。这需要情态动词来完成，而其中使用频率最高的非"shall"和"may"莫属。

（1）"shall"在英语商务合同中使用频率非常高，这主要是因为该词所具有的含义十分广泛。"shall"在一般英语语法中多与第一人称 I/We 连用，表达将来时的动作，在合同文本中可用来表示"义务、许可、命令、责任"等含义，一般表

示当事人的义务,即应做什么。如果对方未履行,则视为违约,并且要承担违约责任或赔偿责任。因此在合同翻译中,"shall"一般翻译成"得、应该或必须"。

【例4】原文:The employer <u>shall</u> have the right to adjust the scope of the works, but <u>shall</u> notify the contractor in writing, and the contract <u>shall</u> accept the project unconditionally.

译文:发包方有权适当调整工程范围,但应书面通知承包方,承包方应无条件接受。

【例5】原文:The contractor <u>shall</u> carry out all the works in accordance with the terms, conditions, and requirements of the contract as defined in the contract terms and to the approval of employer.

译文:承包方应按照有关合同条款中规定的条款、条件及要求进行施工,并由发包方进行认可。

(2)"may"在普通英语中表示"可以",表示主体有一定的选择空间;在法律英语中通常表示法律、法规提出的要求不带有强制性,当事人可以自由决定做或者不做某些行为,或者在一定的条件下允诺或许可做或者不做某些行为,在汉语合同文本中可译成"可以""允许"等。

【例6】原文:Employer <u>may</u> from time to time delegate to the engineer any of the duties and authorities vested in employer as specified in contract and <u>may</u> any time revoke such delegation.

译文:发包方可随时委派工程师,行使合同约定的自己的职权,并可在认为有必要时撤回委派。

(3)shall/may not 是 shall 和 may 的否定形式,常用来表示禁止性的义务,即不能做什么。"shall not"的语气比"may not"稍强。

【例7】原文:The contract <u>shall not</u> be modified or altered unless otherwise agreed in writing by both parties.

译文:若无合同当事双方书面同意,该合同不得进行修改或改动。

3. 正式表达的使用及翻译

工程合同语言属于法律语言的一个分支,因此也具有法律语言的基本特征。这种文体属于"庄重文体",是各种英语文体中规范程度最高的一种。与普通语言相比,工程合同语言更加正式、严肃,逻辑更加严密。因此,合同当中经常使用大量的正式用语。如：

日常英语	合同英语
before	prior to
stop	expire/cease
think	deem
because of	by virtue of
other matters/events	miscellaneous
begin	commence
according to	in accordance with
end	terminate
explain	construe

【例8】原文：Any such delegation or revocation in writing shall, <u>prior to</u> 7(seven) days, be given to contractor. Notification of such delegation or revocation shall be attached to this contract.

译文：委派和撤回均应提前7(七)天以书面形式通知承包方。委派书和撤回通知作为本合同附件。

【例9】原文：Written agreement(s) or document(s) in relation to negotiation/consultation, modification of works between employer and contractor shall be <u>deemed</u> as parts of contract.

译文：发包方与承包方间有关工程的洽商、变更等书面协议或文件应视为本合同的组成部分。

4. 成对词的使用及翻译

一般来说,成对词是指由两个词构成的词组,一般由 and 或 or 连接。其中以两个同义词或近义词组成的词组居多,用两个词来表达同一个含义。合同中经常使用成对词,使合同周密严谨,概念表达精确无误,以减少漏洞和争议。如：

【例10】原文：The contract shall not be <u>modified or altered</u> unless otherwise agreed in writing by both parties.

译文：若无合同当事双方书面同意,该合同不得进行修改或改动。

【例11】原文：The contractor shall assume and perform hereunder all the <u>obliga-</u>

tions and the <u>liabilities</u> of the employer under the main contract in relation to the works.

译文:承包方应在主合同条款下,承担并履行与本工程相关的对发包方的义务及责任。

类似的成对词还有:

terms and conditions 条款

influence or affect 影响

perform and fulfill 履行

null and void 无效

make and enter into 达成

right and interest 权益

by and between 经由

sign and issue 签发

一般情况下,近义词的并列成分中,如两个词没有意义上的差别,只需要译出其中的一个意思,合同的准确严谨性决定了合同中不允许出现哪怕是最细小的失误或是遗漏。不难看出,这种词语并列的用法是工程合同英语准确严谨的又一体现,反映了合同英语对词义准确、文意确切的追求,可以有效地避免诉讼时双方律师利用词义差别来争论。

二、句法层面

1. 被动语态的使用及翻译

工程合同是一种规定各方权利及义务的专业性文书,且经常涉及巨大的经济利益往来,在文字叙述方面要求客观公正及措辞严谨,所以较多使用被动语态。而在汉语合同当中,一般情况下会采用主动语态,被动句往往翻译成无主句或者"的"字结构,避免说出相应的施动者,进而确保语气客观性及公正性。基于英汉两种语言所存在的差异,在对英文合同进行被动语态翻译的过程中,译者需要把被动语态有效转化为主动语态,可以适当地采用无主句以及"的"字结构,从而满足汉语合同的要求。

【例 12】原文:The amount of such costs, which <u>shall be added</u> to the contract price.

译文：应在合同价中增加此类费用总额。

【例13】原文：In the circumstance that project site is constructed by several companies，employer shall be entitled to organize coordination seminar and allocation meeting which shall be attended by contractor(s) on time.

译文：在现场有几个承包单位同时施工的情况下，发包方有权组织协调会，承包方应按时参加。

2.长句的使用及翻译

拟定工程合同时必须确保内容清晰准确、措辞严谨得体，不能出现表述含糊而引起争议的句子。长句和复杂句的使用往往能使合同语义表达清晰准确、细致严密，有效避免双方产生误解与歧义。长句句子结构相对复杂，涉及不少从句和修饰语的使用，也使合同语言相对晦涩难懂。在翻译这些长句时，首先要正确理解各种相关成分的逻辑关系；然后再适当切分，分清句子的主干成分；最后再按汉语的表达习惯，变动语序，重新组合。这样才能连贯、准确、清晰地予以表达。

【例14】原文：The contractor hereby acknowledges that any breach by him of the contract may result in the employer committing breaches of and becoming liable in damages under the Main contract and other contracts made by him in connection with the works and may incur further loss or expense to the employer in connection with the works.

译文：承包方应承认其任何违背合同的行为均可引起发包方违背主合同并对合同项下工程及其他由发包方签订的合同项下的有关工程的毁损负有责任，并可导致发包方产生与工程有关的进一步损失或开支。

在上例中，句子结构较为复杂，包含了宾语从句、后置定语及一系列介词短语。句子主干为"the contractor acknowledges that…"，宾语从句结构为"any breach may…and may…"。在翻译时，应理出句子主干，按照汉语表达习惯进行准确的表达。

三、篇章层面

工程合同是一种措辞严谨的法律文件，必须词义准确、句子严密，不允许合同当事人利用文辞和句义的不确定性和模糊性来逃避责任。程式化的语篇模式已被广泛运用到各种国际商务合同中,以体现其较强的功能性和目的性。工程英语合同一般具有统一规范的格式,语言正式、严谨,篇章结构完整,程式化色彩较强,已形成了一种完整、规范的文体结构特征。合同语篇措辞严谨、结构清晰、语意明确,是专业化、标准化以及法律化的文件。如经常出现"include but not limited to...","and/or"等表达方式来进行范围限定,以避免相关内容成为今后合同双方争议的焦点。

【例15】原文:If the contractor suffers delay and/or incurs costs from failure on the part of the employer to give possession in accordance with the terms, the engineer shall, after due consultation with the employer and the contractor, determine...

译文:如果由于业主未按条款规定让承包方移交现场,使承包方延误了工期和(或)招致费用,则工程师应在及时与发包方和承包方协商之后,决定……

【例16】原文:The scope of work herein includes but not limited to: coal handing system construction and subsidiary facilities foundation, coal yard foundation, warehouse and maintenance house, ESP control house and air compressor room.

译文:工作范围包括但不限于:输煤系统建设以及附属设备基础、煤场基础、材料库及检修楼、电除尘配电室与空压机房的全部建筑工程。

第五节 拓展延伸

一、中译英

(一)甲方义务

1. 为乙方配备电脑及打印机设备、纸张及其他办公耗材等。

2. 为乙方现场资料员提供住宿。

3. 乙方外出到相关单位办事或者送检材料时,为乙方提供交通方便。

(二)乙方义务

1. 至少配备1名具备资质的资料员驻守工地现场办公,乙方指定驻地资料员。驻工地资料员应当相对稳定,未经甲方许可不得擅自更换或撤离资料员。

2. 保证所有资料、半成品试验及时到位,与工程进度同步进行,不得因材料实验、资料的原因影响材料的使用及工程进度。

3. 认真落实资料收集整理、归档的相关制度,按照本合同的约定完善资料的编制整理。

4. 及时传达政府职能部门的有关指示、指令,负责与检测中心及相关管理单位的联络及相关工作。

5. 负责资料的编制、整理及其他专业分包单位资料的收集、汇总整理及资料的备案工作。

6. 按时参加甲方、建设方、监理单位召开的各种例会、技术方案会、施工方案会,并整理好会议纪要,转发给各建设参与单位。

7. 配合工程建设各分包单位办理签证,及时传达管理单位对各施工单位的工作联系函及处罚指令。

8. 乙方工作人员应服从甲方管理,因乙方原因发生的所有事故,由乙方承当责任,与甲方无关。

9. 乙方有对工程资料进行保密的义务。

(三)违约责任

1. 甲方应按约定时间和数量向乙方支付资料款,如不能按约定时间付款,乙方有权要求甲方赔偿由此造成的全部损失。

2. 乙方资料员不得随意擅自离开工地现场,如需离开应向甲方请假,如未请假离开,每次罚款 200 元。

3. 如乙方未按规范规定做齐所有相关实验或存在弄虚作假的行为,一经发现,甲方有权给予经济处罚。实际每项少做或少送一次样品实验,按实验室单项试验费双倍扣回并处以 500 元/次的罚款,并视问题的严重程度保留追究乙方责任的权力。

4. 乙方应对所有资料的时效性、准确性、完整性负责。若乙方资料不准确、及时、完整,影响甲方工程进度或者工程验收,甲方有权要求乙方支付违约金。若该违约金不足以弥补由此给甲方造成的损失,乙方还应当赔偿由此给甲方造成的全部损失。

(四)合同变更和终止

1. 除出现本合同约定解除的情形外,未经双方协商一致,任何一方均不得擅自变更或者解除本合同。

2. 乙方应当及时满足甲方在监督管理过程中提出的整改要求,若乙方未在甲方要求的合理期限内有效整改,甲方有权单方解除本协议。

3. 工程竣工验收、资料备案完善,付清工程资料费用后,本合同书自动终止。

二、英译中

1. GENERAL

1.1 All the safety and quality requirements provided by the first party are to be strictly followed by the second party.

1.2 The second party shall complete all the works listed in contract scope of work as per the requirements stipulated in the drawings and specifications provided by the first party.

1.3 The second party undertakes to use the competent personnel, qualified machines, measuring devices of any kind needed for construction to ensure that each of the final works meets the acceptance criteria stipulated in the relevant drawings and specifications.

1.4 The second party shall be not, without the first party' written consent, to subcontract this works to the third party. Otherwise, without prejudice to any other rights of the first party, the second party shall bear any and all loss and responsibility resulting from its default.

2. RESPONSIBILITIES

2.1 First Party

2.1.1 Monitoring and controlling the project progress, quality and safety.

2.1.2 Obtaining the visa required for the second party's personnel entering into Saudi Arabia and permits to construction site and ensuring the second party's personnel access into site as planned. The second party shall cooperate and provide the relevant documents. In case of the delay in obtaining the above mentioned visas and permits not attributable to the second Party and consequently significant influence on the fulfillment of milestone dates set forth in the subcontract, the parties shall discuss and solve the issue and the first party shall grant the second party appropriate time extension of the milestone dates.

2.1.3 Providing as planed the complete technical documents, which including AFC drawings, document format, relevant specifications and list of applicable standards to ensure the construction progress not hindered by the reason thereof.

2.2 Second Party

2.2.1 Being monitored and controlled by the first party of construction progress, work quality and construction safety.

2.2.2 Submitting work progress reports on time to the first party, which including daily report, weekly report and monthly report.

2.2.3 Performing the work as per the construction schedule and requirements set forth in the drawings and specifications to meet the acceptance criteria stipulated in the contract.

2.2.4 Obeying the instruction of the first party for mobilization as per the construction schedule and arrangement.

2.2.5 Taking appropriate actions to maintain and safeguard the completed works until completion acceptance. The first party shall arrange for the acceptance in time upon the application for acceptance from the second party.

2.2.6 Submitting quality control dossier including quality certificates all of materials provided by the second party to the first party for approval, applicable for the second party supplied materials only.

2.2.7 Preparing the construction records and as-built documents as per the document format provided by the first party, and submitting to the first party for approval.

3. TIMING OF WORK

3.1 The second party will execute the works in accordance with appendix, this schedule may be modified if it becomes convenient for planning or production necessities.

3.2 The second party shall schedule and coordinate carrying out of the work to meet the schedule requirements set forth in scope of work of this contract. The second party shall submit to the first party for approval, a detailed schedule showing the sequence in which the second party proposes to perform the work, the start and completion dates of all separable portions of the work, manpower forecasts, materials procurement and delivery plans and any other information specified by the first party. The second party will adhere to the schedule approved by the first party and attend scheduled progress and coordination meetings called by the first party.

3. 3 During the performance of work, the second party shall submit to the first party periodic progress reports on the actual progress and updated schedules as may be required by this contract or requested by the first party. If the second party's performance of the work is not in compliance with the schedule established for such performance, the first party may, in writing, require the second party to submit its plan for schedule recovery, or specify in writing the steps to be taken to achieve compliance with such schedule, and/or exercise any other remedies under this contract. The second party shall thereupon take such steps as directed by the first party or otherwise necessary to improve its progress without additional cost to the first party.

3. 4 The second party recognizes that the first party, other contractors and the second party may be working concurrently at the jobsite. The second party shall cooperate with the first party and other contractors so that the project as a whole will progress with a minimum of delays. The first party reserves the right to direct the second party to schedule the order of performance of its work in such manner as not to interfere with the performance of others.

4. PERFORMANCE SECURITY

4. 1 The second party shall, within 30 (thirty) days after the signature date of the subcontract, procure for the benefit of the first party an approved, unconditional and irrevocable on-demand bank guarantee as a PERFORMANCE GUARANTEE in the sum of 10% (ten percent) of the estimated total subcontract price from a bank which is acceptable to the first party in his absolute discretion. The second party shall bear all costs and expenses incurred in obtaining such guarantee and the same shall be deemed to be included in the subcontract price.

4. 2 The PERFORMANCE GUARANTEE shall be maintained in full force and effect until the subcontract work being completed and accepted by the first party.

5. INSPECTION

5.1 The first party is entitled to inspect and supervise all construction activities at site.

5.2 Any part of work which does not pass the inspection or test, after being qualified as defective shall be corrected immediately at the cost of the second party.

5.3 As to other quality related issue not attributable to the second party, it shall be settled mutually greed.

6. SUSPENSION

The first party shall have the right to suspend the contract work if the quality of the works does not meet the requirements set forth in the contract.

7. FORCE MAJEURE

7.1 Force majeure includes but not limit to terrorism, war, riot, flying object falling or such natural disasters as explosion, fire, which not attributable to the first party or the second party, as well as flood, typhoon, earthquake, infectious disease, epidemic etc.

7.2 Under such circumstances, the second party shall advise the first party within 72(seventy-two) hours and take all necessary measures to minimize the impact on the performance of the work under the assistance of the first party.

7.3 In case the accident lasts for more than 8 weeks, the first party shall have the right to cancel the contract.

8. SETTLEMENT OF DISPUTES

8.1 Any dispute between the two parties shall be settled amicably.

8.2 If amicable settlement cannot be reached, it shall be resolved on the mutually agreed court in accordance with the laws of China.

第五章 工程技术文件

工程技术文件的编制、收集、整理是工程施工过程管理中的一项重要内容。它不仅是工程的重要档案资料、工程建设实际情况的反映,也是工程建成后运行、维护必不可少的依据。

第一节 背景分析

随着全球化步伐的加快和现代工程技术的不断发展,工程技术发展水平越来越高,工程技术文件的作用也越来越重要。工程技术文件包括在整个工程施工过程中形成的、具有归档保存价值的各种技术文件材料,包括从工程项目开工到竣工全过程形成的文字材料、图纸、图表、计算材料、照片、录像资料、磁盘等,特别是与业主和监理工程师的往来文件、施工记录、技术交底资料、设计变更文件、竣工图和竣工验收文件等。

一、工程技术文件的种类

1. 工程设计文件

工程设计文件是工程建设的依据和基础,包括初步设计、施工图设计、工程招标文件等。

2. 施工文件

施工文件是工程建设实施的依据和标准,包括施工计划、安全生产方案、质量控制计划等。

3. 验收文件

验收文件是工程建设验收的依据和标准,包括验收方案、验收资料、验收记录等。

4.变更文件

变更文件是工程建设在实施过程中对原设计的修改和调整,包括设计变更、施工方案变更、验收标准变更等。

5.维护文件

维护文件是对工程建设建成后的维护和保养,包括设备安装调试记录、设备维护保养记录等。

二、工程技术文件的重要性

首先,工程技术文件是工程建设实施过程中的重要依据,对确保工程质量和进度有着重要的作用。同时在工程建设前后起着重要的指导作用,可以解决和避免工程建设过程中的问题,通过合理的设计、科学的施工方案、准确的验收标准等,确保工程建设的质量和进度。其次,工程技术文件是工程建设管理的重要依据。通过对工程建设过程中的工作进行全面的记录和资料汇总,可以提高工作效率和管理水平。工程技术文件在现代工程建设中不可或缺,是确保工程建设质量和进度的必要手段,对于每一个参与工程建设的人员都具有重要的意义。

三、工程技术英语的特点及翻译

工程技术即在工业生产中实际应用的技术,也就是说,人们将科学知识或利用技术发展的研究成果应用于工业生产过程,以达到改造自然的预定目的之手段和方法。工程技术英语(English for engineering and technology)就是用英语表达或传达工程项目建设或者技术实践活动的专业语言,已经成为工程技术人员进行对外技术交流的重要方式。与其他文体相比,工程技术英语具有自身的特点:

1.频繁使用专业词语

专业性是工程技术英语的主要特征,因此专业词语在工程英语中所占的比重非常大。工程技术英语涉及的学科很广泛,诸如建筑、矿业、机械、电气、船舶、地质等,这些学科的术语对翻译也造成了一定的困难。在翻译过程中,碰到较难的专业术语时,译者一定要多与客户及时沟通,并邀请专业人士进行审校,以保证工程技术文本译文的准确性。不仅如此,译者还要注意不少专业术语在不同学科领域具有不一样的意义。

术语	不同领域的不同意思
reaction	"反作用力"（物理） "反应堆"（核能） "化学变化"（化学）
power	"做功"（物理） "电力"（电力） "幂"（数学）
carrier	"带菌体"（医学） "载波"（计算机） "刀架、托架"（机械）
amplitude	"波幅"（医学） "幅角"（数学） "振幅"（电子信息）

2. 频繁使用缩略语

在工程技术英语中，缩略语很常见，尤其是工程技术图纸，这是为了简便而采用的一种表达方式。缩略语是科技词汇重要的构词方式，具有简洁明快的特点，提高了语言交流的效率。缩略语大多以首字母缩略（acronym）为主，也有截短缩略（clipping）、拼缀缩略（blend）等形式。

缩略语	全称	译文
ACAS	Airborne Collision Avoidance System	机载防撞系统
OLT	Optical Line Terminal	光线路终端
SLAS	Short Arm/Long Arm Suspension	长短臂悬架
EPB	Electric Parking Brake	电子驻车制动系统
SAS	Steering Angle Sensor	转向角传感器
NAVAID	Navigation Aids	助航设施
TeleSat	Telecommunications Satellite	通信卫星

3. 频繁使用被动语态

工程技术英语中被动语态的使用频率较高，这也是工程技术英语句法上的一个主要特点。据统计，被动语态句子在工程技术文本中占三分之一的比例，比普通英语文章中被动语态句子出现的频率高一倍。被动语态与主动语态相

比,更能突出要说明的事物——动作对象,所以才在技术文章和科普作品中得到广泛使用。如:

【例1】原文:The rock-backfill slinger <u>is fixed</u> under the skip on a turntable in a central position under the hatch and the discharge chute. Thus the feed chute <u>is also maintained</u> in a central position under the hatch during the swiveling and rotating movements so that there is only minimum loss of material while dumping.

译文:在出料口和排料溜槽下方的中心处有一旋转台,而台上料斗下方安装有废石充填抛掷机。因此,在旋转运动中溜料槽总是保持在出料口下方的中心位置,故卸料所产生的物料损失最少。

工程技术英语的特殊性对从事工程技术翻译的译者提出了更高的要求。在翻译过程中,译者千万不能因为理所当然的自我理解而导致译文出现失误。总的来说应该遵循以下三个原则。

1. 准确性原则

工程技术领域的文本内容通常都是非常专业和复杂的,译者要具备丰富的专业知识和语言技能才能准确表达原文的意思。因此,在翻译工程技术文本时,译者必须仔细阅读原文,理解其含义,还要对译文进行反复校对和修改,确保翻译结果的准确性和可信度。

2. 一致性原则

在翻译过程中,译者需要尽可能地保持原文的语言风格和用词习惯,以确保译文的一致性。在翻译术语和专业名词时,译者需要遵循一定的规范和标准,以确保术语的一致性和准确性。

3. 流畅性原则

技术文本通常都是非常专业和复杂的,译者需要具备良好的语言表达能力和翻译技巧,以确保译文的流畅性和可读性。在翻译过程中,译者需要注意词语搭配和语法结构,以确保译文语言表达准确、简洁、清晰,并符合读者的阅读习惯。

第二节　专业术语

序号	中文	英文
1	岸边溢洪道	river-bank spillway
2	坝顶	dam crest
3	坝顶宽度	crest width
4	坝高	dam height
5	坝基渗漏	leakage of dam foundation
6	坝肩	dam abutment
7	坝坡	dam slope
8	坝轴线	dam axis
9	薄壁堰	sharp-crested weir
10	薄拱坝	thin-arch dam
11	饱和电流	saturation current
12	并联电阻	shunt resistance
13	参数	parameter
14	操纵基因	operator gene
15	侧槽溢洪道	side channel spillway
16	柴油发动机	diesel engine
17	铲斗	scoop
18	铲土机	earth scraper
19	常压潜水服	normobaric diving suit
20	沉降	settlement
21	沉沙池	sediment basin
22	沉沙条渠	sedimentary channel
23	承载能力	bearing capacity
24	齿墙	key wall
25	冲沙闸	flush sluice
26	充电桩	charging pile

续表

序号	中文	英文
27	抽排措施	pump drainage measure
28	传感器	sensor
29	船闸	navigation lock
30	串联电阻	series resistance
31	单晶硅	monocrystalline silicon
32	单线船闸	single-line lock
33	导航系统	navigation system
34	导流洞	diversion tunnel
35	倒虹吸管	inverted siphon
36	地基变形	foundation deformation
37	地下水	groundwater
38	地形	terrain
39	地形摄影测量	terrestrial photogrammetry
40	地形学	topography
41	电动泵	electrical pump
42	电锯	power saw
43	电气导线	electrical conductor
44	吊车	crane
45	吊管带	pipe sling
46	调节基因	regulatory gene
47	丁坝	spur dike
48	丁钢条	T steer bar
49	丁字梁	T-beam
50	定向仪	orientation device
51	动水压力	hydrodynamic pressure
52	洞轴线	tunnel axis
53	断路器	circuit breaker
54	多晶硅	polycrystalline silicon

续表

序号	中文	英文
55	多线船闸	multi-line lock
56	额定功率	rated power
57	繁殖系统	breeding system
58	反铲挖掘机	back digger
59	非晶硅	amorphous silicon
60	分子生物学	molecular biology
61	浮式采油	floating production
62	辐射量	radiation quantity
63	辐射强度	radiation intensity
64	干式变压器	dry-type transformer
65	刚性管道	rigid pipeline
66	钢筋计	reinforcement meter
67	高压喷射装置	high pressure jets
68	隔热层	insulation layer
69	隔水管连接器	riser connector
70	工程测量	engineering survey
71	功率因数	power factor
72	固结灌浆	consolidation grouting
73	管道材料	pipe material
74	管道阀	pipeline valve
75	光电转换	photoelectric conversion
76	光伏	photovoltaic
77	光伏电缆	PV cable
78	光伏方阵	PV matrix
79	光纤电缆	fiber optic cable
80	海底井口	seafloor wellhead
81	海底水位	sea-bed level
82	海流速度	current velocity

续表

序号	中文	英文
83	海上生产平台	offshore production platform
84	海上油气管道	offshore hydrocarbon pipeline
85	海生物聚集	marine growth
86	海洋隔水管	marine riser
87	焊工	welder
88	夯土机	tamper
89	虹吸溢洪道	siphon spillway
90	互感器	transformer
91	回填灌浆	backfill grouting
92	回填压实机	backfill compactor
93	汇流箱	combiner box
94	混凝土试力砖	test cube
95	混凝土重力坝	concrete gravity dam
96	机械手臂	manipulator arm
97	基因工程	genetic engineering
98	基因克隆	gene cloning
99	加热电缆	heating cable
100	剪力墙结构	shear wall structure
101	剪切应力	shear stress
102	交流	AC
103	搅拌机	stirring machine
104	接地网	grounding grid
105	接口	interface
106	结构系数	structural coefficient
107	截水槽	cutoff trench
108	晶体缺陷	crystal defect
109	井式溢洪道	shaft spillway
110	静水压力	hydrostatic pressure

序号	中文	英文
111	抗冲刷性	scour resistance
112	抗冻性	frost resistance
113	抗拉强度	tensile strength
114	抗生素	antibiotic
115	抗原性	antigenicity
116	科里奥利力	Coriolis force
117	可焊性	weldability
118	空腹拱坝	hollow arch dam
119	空腹重力坝	hollow gravity dam
120	库区	reservoir area
121	宽顶堰	broad crested weir
122	沥青路	tarmac road
123	路面破碎机	road crusher
124	轮式装载机	wheel loader
125	螺纹栓	threaded bolt
126	滤波器	filter
127	木模板	timber formwork
128	耐腐蚀性	corrosion resistance
129	耐火性	fire resistance
130	泥沙压力	silt pressure
131	逆变器	inverter
132	排水	drainage
133	配电箱	distribution box
134	平路机	grader
135	破碎机	crushing machine
136	铺瓦范围	tile coverage
137	启闭机	hoist
138	牵引拖拉机	tow tractor

续表

序号	中文	英文
139	倾斜仪	clinometer
140	渠首	canal head
141	热处理	heating treatment
142	熔断器	fuse
143	扇形闸门	sector gate
144	设计容量	design capacity
145	深水管道	deepwater pipeline
146	渗流	seepage
147	生殖细胞	germ cell
148	石油管道运输系统	oil pipeline transportation system
149	输出电压	output voltage
150	水工隧洞	hydraulic tunnel
151	水合物形成	hydrate formation
152	水利枢纽	hydraulic complex
153	水磨石砖	terrazzo tile
154	水平面	horizontal plane
155	水下机器人	remote operated vehicle
156	水压力	hydraulic pressure
157	松土机	ripper
158	锁坝	closure dike
159	拓扑	topology
160	碳水化合物	carbohydrate
161	填土	fill
162	通信接口	communication interface
163	同系交配	inbreeding
164	同源重组	homologous recombinant
165	同种的	homogenic
166	土石坝	earth-rockfill dam

续表

序号	中文	英文
167	推力承座	thrust block
168	推土机	bulldozer
169	拖车	trailer
170	拖杆	tow bar
171	挖沟机	ditch excavator
172	外部静水压力	external hydrostatic pressure
173	尾水渠	tailwater channel
174	卫星井	satellite well
175	温控系统	temperature control system
176	系梁	tie beam
177	细胞周期	cell cycle
178	箱式变压器	box-type transformer
179	消力池	stilling basin
180	雄榫接	tenon joint
181	悬臂梁	cantilever beam
182	阳极的	anodic
183	液压管	hydraulic tube
184	液压管线	hydraulic line
185	遗传学	genetics
186	异种的	heterogenic
187	引水渠	diversion canal
188	营养细胞	vegetative cell
189	油田开发	oilfield development
190	油田寿命	oilfield life
191	远系繁殖	outbreeding
192	杂交	hybridization
193	黏性土	cohesive soil
194	振动夯实机	vibrating tamper

续表

序号	中文	英文
195	支护桩	tangent pile
196	直流	DC
197	指标	index
198	制动器	stopper
199	终端沙井	terminal manhole
200	重组 DNA	recombinant DNA
201	转移 RNA	transfer RNA
202	自体受精	self-fertilization
203	阻尼比	damping ratio
204	组态	configuration
205	钻杆柱	drill pipe string

第三节　翻译案例

案例一：英译中

原文：

SAR500-A High-Precision High-Stability Butterfly Gyroscope
with North Seeking Capability(Excerpt)

There is an ever-increasing need for high-precision, high-stability and miniaturized MEMS angular rate sensors for inertial navigation, positioning, stabilization, tracking and guidance of remote operated devices. In the market of tactical and inertial grade MEMS gyroscopes, proving the reliability and long-term stability of these devices remains probably the greatest challenge.

SAR500, a robust tuning fork type MEMS gyroscope with SPI communication, has been designed for vibration-exposed applications operating in harsh environments. For operating as angular rate sensors, the tuning fork type gyroscopes require arrangements with at least two orthogonal degrees of freedom. In such devices, a certain known primary motion, corresponding to the first degree of freedom, must be generated and maintained in the sensor. An external angular velocity affecting the sensor in a direction perpendicular to the primary motion induces an oscillating Coriolis force in a direction corresponding to the second degree of freedom. The induced Coriolis force is proportional with the external angular velocity and the amplitude of the primary motion. It is therefore necessary to generate, maintain and control a primary oscillation with large amplitude, achievable only by structures sealed in high vacuum.

Commonly, in silicon-based tuning fork gyroscopes, the primary motion is initiated and maintained at right angles to the substrate surface by means of electrostatic excitation. However, the rather low oscillation amplitude will cause these types of gyroscopes to suffer from lower gyroscopic scale factors. This problem has been addressed by using beams that have a tendency to bend in a direction that is substantially parallel to the plane of the substrate, thus allowing primary motions with large oscillation amplitudes.

A limitation in many existing devices is the presence of non-uniform characteristics and built-in stress that can cause unintended sensitivity to external mechanical and thermal loads. This problem has been previously addressed mainly by use of stress-release structures and pedestals. In order to further improve the uniformity of the mechanical characteristics and the long-term stability of the gyroscope, a new fabrication method has been developed. This method allows the use of single-crystal silicon in the entire structure of the device, including the capping wafers, while enabling the hermetic sealing of the vibrating elements by means of anodic bonding.

Closed feed-back loops are used to control the excitation and detection modes. Furthermore, SAR500 uses additional electrodes in order to continuously adjust the frequency of the oscillations and actively compensate the quadrature bias.

The SAR500 contains a Butterfly Gyro MEMS die and an analog ASIC, individually housed in rigid, fully symmetrical, custom-made ceramic packages with high thermal efficiency, which further improves the device long-term stability. Digital ASIC, or an FPGA, contains the needed control and functional algorithms to achieve the superior performance.

The sensing element consists of two identical seismic masses suspended by means of asymmetric driving beams to pedestals designed to minimize the mechanical and thermal stress. The two masses are connected to each other by means of a centrally located synchronizing beam. The sensing principle is an acceleration-sensitive resonant structure, with an ASIC for resonance control and signal conditioning.

The seismic masses and the beams are arranged to provide a first degree of rotational freedom about the excitation axes, which are perpendicular to the plane of the substrate, and a second degree of rotational freedom about the detection axis, which is coincident with the longitudinal axis of the beams. The neutral axis of the beams forms an acute angle α with the normal excitation axes. The beams have thus the tendency to bend in a direction that is mostly parallel to the plane of the substrate, so that an in-plane oscillation of the masses may be initiated by a force that is perpendicular to the plane of the substrate.

The driving beams are dimensioned in such a way that the resonance frequency of the in-plane bending mode matches the resonance frequency of the torsion mode. Henceforward, the in-plane bending mode of the springs will be referred as "primary" or "excitation" mode, while the torsion mode will be referred as "secondary" or "detection" mode.

Capacitive schemes, operating in closed feedback loops, are employed to initiate, control and accurately quantify the primary and secondary motions. Recesses are etched on both sides of the silicon substrate to form the gaps of the capacitors.

The vibrating structure, formed by the two seismic masses and the beams, is attached to and hermetically sealed between two, fully symmetrical capping silicon-glass composite wafers. A silicon frame, manufactured in the same single-crystal silicon substrate, surrounds the vibrating structure. The frame and pedestals are attached to the silicon-glass composite wafers by means of anodic bonding in high vacuum.

译文：

SAR500——具有寻北能力的高精度、高稳定性蝶形陀螺仪(节选)

高精度、高稳定性的小型微机械(MEMS)角速率传感器可用于远程操作设备的惯性导航、定位、稳定、跟踪和引导，而且此方面的需求日益增加。在战术级和惯性级 MEMS 陀螺仪市场上，证明这些设备的可靠性和长期稳定性仍然是最大的挑战。

SAR500 是一款强大的音叉型 MEMS 陀螺仪，具有串行外设接口(SPI)通信功能，专为在恶劣环境中的振动暴露应用而设计。音叉型陀螺仪用作角速度传感器，需要至少两个正交自由度。在这种设备中，传感器中必须产生并维持与第一自由度相对应的某种已知主运动。外部角速度在垂直于主运动的方向上影响传感器，在与第二自由度相对应的方向上引起振荡的科里奥利力。诱导产生的科里奥利力与外部角速度和主运动幅度成正比。因此通过高真空中密封结构产生维持和控制大幅度的主振荡是有必要的。

通常，在基于硅的音叉型陀螺仪中，主运动通过静电励磁启动并与衬底表面保持垂直。然而，较小的振荡幅度将导致该类型陀螺仪受到较低陀螺仪比例因子的影响。该问题已通过使用横梁得以解决，这些横梁倾向于朝平行于衬底平面的方向弯曲，问题解决之后产生大幅度振荡的主运动。

许多现有设备的局限性是存在不均匀性和内在应力，这可能导致对外部机械负荷和热负荷不可预知的敏感性。在过去，主要通过使用应力释放结构和衬底来解决该问题。为了进一步提高陀螺仪的均匀性和长期稳定性这两个机械特性，人们开发了一种新的工艺方法，该方法允许在包括盖帽晶圆的整个设备结构中使用单晶硅，同时通过阳极键合达到振动元件的气密密封。

闭合反馈回路用于控制励磁和检测模态。此外，SAR500 使用附加电极来连续调节振荡频率并主动补偿正交偏置。

SAR500 内含一个蝶形陀螺仪 MEMS 裸片和模拟专用集成电路(ASIC)，独立安置在坚硬且完全对称的定制陶瓷封装中，该封装具有高热效率，进一步提高了设备的长期稳定性。数字专用集成电路(ASIC)或现场可编程门阵列(FP-GA)包含所需的控制算法和功能算法，以实现卓越的性能。

感应元件由两个相同感振质量块组成，通过非对称驱动梁悬挂在基座上，以达到机械应力和热应力最小化，两个质量块通过位于中心的同步梁相互连接。传感是一种加速度敏感的共振结构，使用专用 ASIC 来控制谐振和调节信号。

感振质量块和横梁围绕垂直于基板平面的励磁轴提供第一个旋转自由度，并围绕检测轴提供第二个旋转自由度，与横梁纵轴重合。横梁的中轴与正常的励磁轴形成锐角 α。因此，横梁在大致平行于衬底平面的方向上具有弯曲的倾向，所以可通过垂直于衬底平面的力引起质量的面内振动。

驱动梁设定为面内弯曲模态的共振频率与扭转模态的共振频率相匹配。因此，弹簧的平面内弯曲模态被称为"主模态"或"励磁模态"，而扭转模态则被称为"次模态"或"检测模态"。

采用在闭合反馈回路中的电容方案来启动、控制和精确量化主运动和次运动。同时在硅衬底的两侧蚀刻凹槽，以形成电容器的间隙。由两个感振质量块和横梁形成的振动结构连接并气密封在两个完全对称的封盖硅玻璃复合晶圆之间。在相同的单晶硅衬底中制作硅框架来包围振动结构，框架和基座通过高真空下的阳极键合附着到硅玻璃复合晶圆上。

案例二：中译英

原文：

分布式光伏发电系统(节选)

能源危机、环境污染受到全球关注，仅仅依靠扩大电网规模显然不能解决这些问题，于是分布式发电作为集中式发电的有效补充产生了。它具有污染少、可靠性高、能源利用效率高、安装地点灵活等诸多优点，有效解决了大型集中电网的许多潜在问题。

光伏电池是把太阳的光能直接转化为电能的基本单元，电池通过组合形成电池组件，电池的光伏性能决定了电池组件的发电特性，电池组件是光伏电站的基本发电设备。从第一块光伏电池问世到现在，光伏发电技术不断发展，电池种类众多、性能各异。商用的太阳电池主要有以下几种类型：单晶硅电池、多晶硅电池、非晶硅电池、碲化镉电池、铜铟镓硒电池等。

晶体硅太阳电池包括单晶硅太阳电池、多晶硅太阳电池、带状硅太阳电池、球状多晶硅太阳电池等。单晶硅太阳电池和多晶硅太阳电池以稳定的光伏性能和较高的转换效率，成为光伏发电市场的绝对主流，在世界各地得到了广泛的应用。

单晶硅太阳电池以高纯的单晶硅棒为原料,是当前开发很快的一种太阳电池,它的结构和生产工艺已定型,产品广泛用于空间和地面。为了降低生产成本,现在地面应用的太阳电池大多采用太阳能级的单晶硅棒,材料性能指标有所放宽,也可使用废次硅材料制成太阳电池专用的单晶硅棒。

虽然单晶硅太阳电池转换效率高,但电池片存在倒角,使得有效发电面积减小。单晶硅光伏组件更适合建设场地面积有限而对工程发电功率要求高的发电项目。另外,根据试验室和工程中的测试数据,单晶硅太阳电池功率衰减较多晶硅太阳电池快。

多晶硅太阳电池使用的多晶硅材料,多半是含有大量单晶颗粒的集合体,或用废次单晶硅材料和冶金级硅材料熔化浇铸而成,然后注入石墨铸模中,待慢慢凝固冷却后,即得多晶硅锭。这种硅锭可铸成立方体,以便切片加工成方形太阳电池片,可提高材料的利用率,组装较为方便。同单晶硅太阳电池相比,多晶硅太阳电池转换效率稍低,但单瓦造价相对便宜,尤其是大功率组件价格要更便宜,适合建设项目用地比较充足的工程,而单晶硅太阳电池更适合建设项目用地紧缺、要求更高转换效率的工程。

薄膜太阳电池组件相对晶体硅太阳电池组件而言,转换效率较低,建设占地面积大。我国大陆地区较少有大规模生产碲化镉薄膜太阳电池组件、铜铟镓硒薄膜太阳电池组件的厂商,产品采购主要依赖进口,且其价格比非晶硅薄膜太阳电池组件高。

本工程太阳电池组件的造价在工程造价中的比重相对较高,有必要降低太阳电池组件价格以节省工程投资。综合考虑本工程的建设用地情况,推荐选用大功率多晶硅电池组件。本项目中,光伏场区接地采用金属支架彼此相连、分块集中接地的方式,接地体选择耐腐蚀性的热镀锌扁钢接地体等。本工程所有逆变器均采用负极接地措施,以防止 PID 效应造成衰减问题。同时采用科学系统的方法解决管理不到位引起的耗能问题,提高能源使用效率。

为了使光伏方阵表面接收到更多太阳能量,根据日地运行规律,方阵表面最好朝向赤道(方位角为 0°)安装,并且应该倾斜安装。因此,只要使方阵面上全年接收到最大辐射量即可保证光伏组件发电量最大。

译文：

Distributed PV Power Generation System (excerpt)

The energy crisis and environmental pollution is attracting worldwide attention. Obviously, these problems cannot be solved only depending on expansion of power grid scale, therefore, the distributed generation emerges as an effective supplement to centralized generation. It is characterized by low pollution, high reliability and energy efficiency, flexible installation position and other advantages, thus effectively solving many potential problems of large-scale centralized power grid.

PV cells are basic units that convert light energy of the sun into electrical energy directly. Cells are combined to form cell modules and the PV performance of cells determines power generation characteristics of cell modules. Cell modules are basic power generation equipment of PV power stations. Since the invention of the first PV cell, PV power generation technology has been continuously developed, with many types of cells generated and different properties performed. Commercial solar cells mainly include monocrystalline silicon cells, polycrystalline silicon cells, amorphous silicon cells, cadmium telluride cells, copper indium gallium selenide cells, etc.

Crystalline silicon solar cells include monocrystalline silicon solar cells, polycrystalline silicon solar cells, ribbon silicon solar cells, spherical polycrystalline silicon solar cells, etc. Monocrystalline silicon solar cells and polycrystalline silicon solar cells are the mainstream in the PV power generation market because of stable PV performance and high conversion efficiency, and they have been widely used in the world.

Monocrystalline silicon solar cells are made of highly purified silicon single crystal rods and are currently developing at a high speed. The structure and production process of monocrystalline silicon solar cells have been mature, and the cells are widely used in space and on ground. To reduce production cost, most solar cells currently used for ground application are made of solar-grade silicon single crystal rods. The requirements for material performance standard have been lowered, so it is also possible to recycle waste silicon materials to produce silicon single crystal rods special for solar cells.

Although the conversion efficiency of monocrystalline silicon solar cells is high, cell sheets are chamfered, so that the effective power generation area is reduced. Monocrystalline silicon PV modules are more suitable for power generation projects which have limited construction area and high requirements for generated power. In addition, according to laboratory and project test data, the power attenuation of monocrystalline silicon solar cells is faster than that of polycrystalline silicon solar cells.

Polycrystalline silicon materials used in polycrystalline silicon solar cells are mostly aggregates containing a large number of single crystal particles. Some of the polycrystalline silicon materials are made by melting and casting of waste monocrystalline silicon materials and metallurgical grade silicon materials, and then injected into graphite casting dies. After they are solidified and cooled off, polycrystalline silicon ingots are obtained. These silicon ingots can be cast into cubes so that they can be sliced and processed into square solar cells. In this way, the material utilization rate of is improved and assembly becomes easier. Compared with monocrystalline silicon solar cells, polycrystalline silicon solar cells are lower in conversion efficiency but cheaper in one watt cost especially in the cost of high power modules, so they are applicable to the project with sufficient construction area. However, monocrystalline silicon solar cells are more suitable for the project with scarce construction area and higher requirement of conversion efficiency.

Compared with the crystalline silicon solar cell module, the thin film solar cell module has lower conversion efficiency and requires larger construction area. Since there is rare manufacturer that produces cadmium telluride thin film solar cell modules and copper indium gallium selenium thin film solar cell modules in large scale in the mainland of China, the products purchasing mainly relies on imports, and the price is higher than that of amorphous silicon thin film solar cell modules.

The cost of solar cell modules in the project accounts for a relatively high proportion of the project cost, so it is necessary to reduce the price of solar cell modules to save the project investment. By comprehensively considering the construction site of the project, the high power polycrystalline silicon solar cell modules are recommended. In this project, the grounding of the PV site is performed through interconnection with metal supports in a concentrated grounding way in different blocks. All inverters in the project adopt negative ground in order to prevent the attenuation caused by PID (potential induced degradation) effect. At the same time, scientific and systematic approaches are adopted to solve the energy consumption resulting from improper management, which improves energy efficiency.

To ensure that the surface of PV array receives more solar energy, the surface shall be preferably installed towards the equator (azimuth angle of 0°) and shall be inclined according to the motion law of sun and earth. Therefore, as long as the surface of PV array receives the maximum radiation throughout the year, maximum power generation of PV module can be ensured.

第四节　翻译评析

一、案例一评析

翻译案例一节选自 *SAR500-A High-Precision High-Stability Butterfly Gyroscope with North Seeking Capability*，从蝶形陀螺仪的设计、工作、特点和制造等方面详细介绍了蝶形陀螺仪。

委托方的相关要求如下：(1)译文语言必须清楚明了，行文流畅达意，避免出现文理不通、逐词死译和生硬晦涩等现象；(2)译文内容需真实准确，不需要太过华丽，译文排版、格式需要与原文保持一致；(3)专业词语翻译应遵循一致性，应避免用词不当而产生理解上的歧义。如有相关专业内容难以理解，可随时与委托方沟通交流。

经过分析可发现，原文在词汇上专业性较强，出现了较多的专业术语、多义词和缩略语。从句法上来看，长句较多，定语从句经常出现，采用一般现在时和被动语态。半技术词语中普遍存在一词多义的现象。

(一)词法层面

1. 科技词语的翻译

工程技术英语中的词语可以分为技术词(technical words)、非技术词(nontechnical words）和半技术词（semi-technical words）

技术词是指某一领域中的专业术语,这些术语在英汉两种语言中往往有单独的词语相互对应,很少出现一词多义及一义多词的现象。本案例中的技术词不在少数:

英文	中文
gyroscope	陀螺仪
Coriolis force	科里奥利力
anodic bonding	阳极键合
oscillations	震荡
quadrature bias	正交偏置
silicon	硅
feedback loops	反馈回路
angular rate sensor	角速度传感器
electrostatic	静电的
wafer	晶圆
scale factor	比例因子
mechanical load	机械负荷

【例1】原文:Furthermore, SAR500 uses additional electrodes in order to continuously adjust the frequency of the oscillations and actively compensate the quadrature bias.

译文:此外,SAR500 使用附加电极来连续调节振荡频率并主动补偿正交偏置。

半技术词又称次技术词,是指依赖语境并较频繁使用的词。这些词由普通词语转化而来,既有普通的含义,也有科技含义,是工程技术文体中应用较广的词。科学思想和技术内容的展开需要依靠半技术词来实现。

Term	作为日常词语	作为科技词语
excitation	激励;刺激	励磁
die	死亡;骰子	裸片;压模
bonding	建立关系;黏合	键合

续表

Term	作为日常词语	作为科技词语
recess	休息;山脉等的凹处	凹槽
capping	覆盖;给……戴帽子	封盖
stopper	阻塞物;瓶塞	制动器

【例2】原文:Anodic bonding is used to seal the silicon <u>die</u> between two glass plates to obtain a mechanically rigid structure, and a controlled and stable atmosphere for the resonating structure.

译文:阳极键合用于密封两块玻璃板之间的硅<u>裸片</u>,以获得机械刚性结构,并为共振结构提供一个稳定和可控的环境。

2. 缩略词的翻译

在工程科技英语中,一些名词、短语、术语或概念会在上下文中多次提及。尤其是由较多专业词语组成的短语,读者可能不太熟悉,频繁使用其全称会增加读者的阅读困难,而这些短语往往有缩写用法,一般是由短语中每个单词的首字母组成的。

缩略词	全称	中文名
ASIC	Application Specific Integrated Circuit	专用集成电路
FPGA	Field-Programmable Gate Array	现场可编程门阵列
MEMS	Micro-Electro Mechanical Systems	微机械系统
SPI	Serial Peripheral Interface	串行外设接口

在翻译此类缩略专业用语时,为了让读者理解无障碍,在该缩略语第一次出现时应将其全称翻译出来;在后文再次出现该缩略语时,可以直接沿用其缩略形式,不必将全称译出来。

【例3】原文:Digital <u>ASIC</u>, or an <u>FPGA</u>, contains the needed control and functional algorithms to achieve the superior performance.

译文:数字专用集成电路(ASIC)或现场可编程门阵列(FPGA)包含所需的控制和功能算法,以实现卓越的性能。

3. 名词化结构的翻译

名词化(nominalization)是指在语法上,其他词性的词,一般是动词和形容词,转化为名词词组的过程。因具有客观性、生动性、直观性和简洁性,名词化在科学、技术等专业领域的解释文本中用得比较多。但是在英译汉时,名词化结构经常译为动词。

【例4】原文:The sensing principle is an acceleration-sensitive resonant struc-ture, with an ASIC for resonance <u>control</u> and signal conditioning.

译文:传感是一种加速度敏感的共振结构,使用专用 ASIC 来控制谐振和调节信号。

【例5】原文:This method allows the <u>use</u> of single-crystal silicon in the entire structure of the device, including the capping wafers.

译文:该方法允许在包括封盖晶圆的整个设备结构中使用单晶硅。

【例6】原文:A limitation in many existing devices is the <u>presence</u> of non-uni-form characteristics and built-in stress.

译文:许多现有设备的局限性是存在不均匀性和内在应力。

(二)句法层面

1. 长句的翻译

英汉两种语言句式结构不同:汉语重意合;而英语重形合,句子往往比较长。这是因为在英语句子中修饰成分和并列成分较多,而且各种短语或从句充当句子成分。工程技术英语亦是如此,长句的语法结构层层交织,盘根错节,会给翻译带来较大的挑战。在很多情况下,长句的翻译会用到拆译法。

【例7】原文:This problem has been addressed by using beams that have a tend-ency to bend in a direction that is substantially parallel to the plane of the substrate, thus allowing primary motions with large oscillation amplitudes.

译文:该问题已通过使用横梁得以解决,这些横梁倾向于朝平行于衬底平面的方向弯曲,问题解决之后产生大幅度振荡的主运动。

本句如果翻译为"该问题已通过使用有朝平行于衬底平面的方向弯曲的倾向的横梁加以解决,因此产生大幅度振荡的主运动",则定语冗长,使得整个译文读起来啰唆拖沓。翻译时可以利用拆译法将上句分成三个小短句,使译文读起来更易懂、更流畅。

2. 被动语态的翻译

与普通英语文本相比,工程科技文本具有简练、客观、准确等鲜明的特点,而这些特点也决定了科技英语中被动语态的大量使用。被动语态在英汉两种语言中存在不小的使用差异，因此译者在对科技英语中的被动语态进行汉译时，不可拘泥于原文，要在深入分析和理解原文的基础上，灵活运用多种翻译方法。在很多情况下，英语中的被动语态被译为汉语中的主动句。

【例8】原文：Recesses <u>are etched</u> on both sides of the silicon substrate to form the gaps of the capacitors.

译文：在硅衬底的两侧蚀刻凹槽，以形成电容器的间隙。

【例9】原文：The vibrating structure, formed by the two seismic masses and the beams, <u>is attached</u> to and <u>hermetically sealed</u> between two, fully symmetrical capping silicon-glass composite wafers.

译文：由两个感振质量块和横梁形成的振动结构连接并气密封在两个完全对称的封盖硅玻璃复合晶圆之间。

（三）篇章层面

工程技术英语主要用来客观地阐明事理和论述问题，语言朴实，极少修饰，注重时间概念、事实和逻辑。一般文学作品中常用的修辞手段，如隐喻、拟人、夸张等，在科技英语中较少见到。因此技术文本具有相似的结构语篇特征，叙述时总是开门见山、直截了当。本案例亦是如此，在叙述时把要表达的主要信息尽量前置，使读者能立即抓住问题的重点。

二、案例二评析

翻译案例二为能源工程技术文本，对比不同太阳电池的性能和适用情况。文本层次清楚，结构紧凑，逻辑严密，在词汇、句法和语篇层次方面均体现了工程技术文本的特点，并用到了各种翻译方法。

（一）词法层面

1. 术语翻译

本案例中同样出现了大量能源工程技术领域的专业术语，如单晶硅电池（monocrystalline silicon cells）、多晶硅电池（polycrystalline silicon cells）、非晶硅薄膜（amorphous silicon thin film）、倒角（chamfer）、逆变器（inverter）、薄膜太阳能电池（thin-film solar cells）、多晶硅锭（polycrystalline silicon ingots）等。

2. 词性转换

词性转换是指在翻译过程中，将原文中的某个词的词性转换成其他词性，其目的是忠实地再现原文信息,使译文既符合语言习惯,又通顺流畅,从而提高译文质量,方便读者接受和理解。在工程技术文本中,汉语倾向于多用动词,叙述呈动态;英语倾向于多用名词,叙述呈静态。所以在汉译英过程中,词性转换是不可避免的。

【例10】原文:第一块光伏电池问世到现在,光伏发电技术不断发展,电池种类众多、性能各异。

译文:Since the invention of the first PV cell, PV power generation technology has been continuously developing, with many types of cells generated and different properties performed.

【例11】原文:仅仅依靠扩大电网规模显然不能解决这些问题。

译文:Obviously, these problems cannot be solved only depending on expansion of power grid scale.

3. 减词译法

由于英汉两种语言在句法和表达方式上存在很大的差异,在译文中按照译入语的表达习惯,增加一些词以便更准确地表达原文所表达的意思,或者省略不符合译入语思维习惯或表达方式的词,使译文更加简练清晰。

在汉语语篇中,我们常常会遇到一些用来表示现象、行为、属性的汉语范畴词,它们一般跟在名词、动词或形容词后面,并不表达实际意义,但却可以使行文完整顺畅、表达具体,常见的有"行为""状况""活动""态度""问题""方面""措施"等。如果英译时将这些汉语范畴词一一直译出来,那么势必会造成英语表达啰唆,因此可考虑将其省译。

【例12】原文:本工程所有逆变器均采用负极接地措施,以防止PID效应造成衰减问题。

译文:All inverters in the project adopt negative ground in order to prevent the attenuation caused by PID (potential induced degradation) effect.

【例13】原文:同时采用科学系统的方法解决管理不到位引起的耗能问题,提高能源使用效率。

译文:At the same time, scientific and systematic approaches are adopted to solve the energy consumption resulting from improper management, which improves energy efficiency.

分析：在上述两例中，表达抽象概念的"措施""问题"均被省译。不仅如此，"能源使用效率"译为"energy efficiency"，省译"utilization"，更简洁明了。

（二）句法层面

1.转态译法

与普通英语文本相比，工程技术文本具有简练、客观等特点，因此被动语态运用得较多；而在汉语中被动语态使用频率较低，多数被动意义隐含在词语或语境中。因此，在工程技术文本汉译英过程中，要充分考虑到中英文被动语态的区别。如：

【例14】原文：为了使光伏方阵表面接收到更多太阳能量，根据日地运行规律，方阵表面最好朝向赤道（方位角为0°）安装，并且应该倾斜安装。因此，只要使方阵面上全年接收到最大辐射量即可保证光伏组件发电量最大。

译文：To ensure that the surface of PV matrix receives more solar energy, the surface shall be preferably installed towards the equator (azimuth angle of 0°) and shall be inclined according to the motion law of sun and earth. Therefore, as long as the surface of matrix receives the maximum radiation throughout the year, maximum power generation of PV module can be ensured.

【例15】原文：多晶硅材料用废次单晶硅材料和冶金级硅材料熔化浇铸而成，然后注入石墨铸模中，待慢慢凝固冷却后，即得多晶硅锭。这种硅锭可铸成立方体，以便切片加工成方形太阳电池片。

译文：Some of the polycrystalline silicon materials are made by melting and casting of waste monocrystalline silicon materials and metallurgical grade silicon materials, and then injected into graphite casting dies. After they are solidified and cooled off, polycrystalline silicon ingots are obtained. These silicon ingots can be cast into cubes so that they can be sliced and processed into square solar cells.

分析：在上述两个例句中，原文都是用主动语态表示被动含义，但是在英文译文中，为了避免使用动作执行者，均不约而同地采用了被动语态，充分显示了工程技术文本的客观性。

2.长句重组

汉语长句中一般包含较多的短句，句子结构也比较简洁。而英语长句中有较多的分词结构和从句。英语强调结构，汉语强调语义。因此，在翻译过程中，应根据目的语的逻辑思维和表达习惯,在充分理解原文的前提下,对句子进行拆分和重组。

【例16】原文:同单晶硅太阳电池相比,多晶硅太阳电池转换效率稍低,但单瓦造价相对便宜,尤其是大功率组件价格要更便宜,适合建设项目用地比较充足的工程,而单晶硅太阳电池更适合建设项目用地紧缺、要求更高转换效率的工程。

译文:Compared with monocrystalline silicon solar cells, polycrystalline silicon solar cells are lower in conversion efficiency but cheaper in one watt cost especially in the cost of high power modules, so they are applicable to the project with sufficient construction area. However, monocrystalline silicon solar cells are more suitable for the project with scarce construction area and higher requirement of conversion efficiency.

分析:在上例中,该句的主要框架为"多晶硅电池适合……工程,单晶硅电池适合……工程"。因此,译文将原句拆分重组并用"however"连接两个句子突出逻辑关系,同时增译"so",阐述多晶硅电池为何适合建设项目用地较充足的工程。

(三)语篇层面

语篇是指由一系列有机联系的句子组成的语言单位,具有一定的主题和特定的交际目的。构成语篇的单词、短语和句子都是按照某种逻辑语义关系排列组合而成的,这种词句之间排列组合的基本规律就是语篇之间的衔接与连贯。英语篇章讲究形式上的严密,经常使用许多衔接和连贯的手段来表达句子之间隐含的逻辑关系。汉语的文本组织模式大多是螺旋形的,汉语句子常通过内在的逻辑关系联系在一起。工程技术英语最具特色的美就是逻辑美,因为科技英语是用来阐述科技事实、概念、原理的,因此逻辑缜密,推导合理,无懈可击。

【例17】原文:能源危机、环境污染受到全球关注,仅仅依靠扩大电网规模显然不能解决这些问题,于是分布式发电作为集中式发电的有效补充产生了。它具有污染少、可靠性高、能源利用效率高、安装地点灵活等诸多优点,有效解决了大型集中电网的许多潜在问题。

译文:The energy crisis and environmental pollution is attracting worldwide attention. Obviously, these problems cannot be solved only depending on expansion of power grid scale, therefore, the distributed generation emerges as an effective supplement to centralized generation. It is characterized by low pollution, high reliability and energy efficiency, flexible installation position and other advantages, thus effectively solving many potential problems of large-scale centralized power grid.

分析:在这个例子中,译文将原文分为两个句子,将"obviously"作为副词放于句首,使得句子联系更紧密,同时增加"therefore"和"thus"等衔接词使句子内部的逻辑关系更清晰。除此之外,利用"it"来进行指示照应,代替前文的"distributed generation"。

第五节 拓展延伸

一、中译英

矿区塌陷是由于地下矿产资源大面积被采空后,顶板岩体脱落,地表产生移动和变形,从而形成塌陷坑和塌陷洼地。煤炭资源开采造成的地面塌陷、村庄房屋损害、道路交通破坏、塌陷坑积水以及地下水污染等,严重影响矿区生态环境,限制了矿产开采行业的发展,极其不利于社会经济的可持续发展。因此,以科学的手段对矿区塌陷进行宏观的、动态的、准确有效的监测具有重要的实际意义。光学卫星遥感影像具有光谱信息丰富、时效性强、信息直观形象的特点,可通过对矿区塌陷坑积水信息提取,间接地了解其空间分布情况。2017 年,有专家利用多时相光学遥感数据监测淮南矿区采煤塌陷积水区的时空变化,研究表明该方法可大范围、周期性获取矿区塌陷分布现状。但光学遥感往往只能提供宏观和半定量成果,不能有效获取塌陷区域的形变场和形变量,导致出现漏判、错判现象。合成孔径雷达干涉测量(InSAR)技术能够全天时、全天候工作,可获取厘米级甚至毫米级地表形变信息。2020 年曾采用小基线集技术(SBAS-InSAR)获取矿区年均形变速率,研究表明矿区沉降区与开采进尺具有良好的一致性。但由于侧视雷达相干距离成像特征,对于地表积水区域和地形起伏大的山区,由于雷达波镜面反射、几何畸变现象,无法获取其所在区域的有效信号。因此,结合 InSAR 和光学遥感有助于对矿区地表塌陷的变化区域、变化速度、变化量进行宏观、快速、准确的监测,可动态监测矿区地表塌陷发展过程和未来发展趋势。选取 Sentinel-1A 雷达卫星数据,基于 SBAS-InSAR 时序形变探测技术,开展矿区地表形变信息提取。同时利用多源多时相光学遥感数据,采用目视解译和归一化差异水体指数技术方法,提取 Landsat7 ETM + 、landsat8 OLI、ASTER、Sentinel-2、GF-2 共 5 种传感器 6 个多时相光学遥感影像的矿区积水信息,间接获取塌陷区的空间变化情况。两者各取所长、相互验证,以期为矿区地表塌陷监测提供一种科学、有效的工作方法。

二、英译中

The Series 4100 gas dispensing systems are used to feed gas under vacuum, at a manually selected or automatically controlled rate, as specified, into a stream of water or process liquid.

The maximum capacity of the gas dispensing system is related to: a) The gas being handled; b) The size and location of the vacuum regulator, which may be furnished as an integral component of the dispenser cabinet or as a separate component for remote mounting; c) The size of the remote mounted ejector which is always furnished as a separate component.

Component Description

1. Vacuum Regulator

The vacuum regulator is of the spring-opposed, diaphragm type which serves, via the throttling action of its vacuum operated inlet valve, to reduce the gas from a varying supply pressure to a constant regulated vacuum for safe transport to the point (s) of application.

Each regulator is equipped with a vacuum actuated, manual reset, 3-position status lever which moves from its operating position, as manually set when the regulator and its associated gas supply system is placed in operation, to its no gas or empty position to provide visual indication that the supply of gas is exhausted or has been interrupted. The other position, the reverse position, is functional only in those instances where the gas dispensing system is of the dual regulator type, serving in such instances to provide for manually locking out the standby regulator and its associated gas supply until the supply of gas associated with the in service regulator is exhausted at which time it moves from its reserve to operating position, thereby providing for an automatic changeover of gas feed from the exhausted in service to the full standby gas supply system. The lever has no functional significance when the regulator is mounted within the dispenser cabinet.

A low temperature switch is offered as an option in the vacuum regulator. The low temperature switch should be connected to the plant alarm system in order to warn of any carry-over of liquid chlorine. It should be specified whenever the source of chlorine gas if from an evaporator.

2. Manual Flow Control

The manual flow control valve, adjustable from the front of the cabinet, is a needle type throttling valve and provides for manual selection of the desired gas flow rate.

3. Automatic Flow Control Valve

The automatic flow control valve is an electromechanical orifice type throttling valve and provides, via its integrally mounted servo type actuator and associated standard control knobs or optional selector switches, for either manual selection of the desired gas flow rate, or automatic regulation of the required gas flow rate in response to the electrical control signal(s).

The standard control knobs, which are integral components of the automatic flow control valve, are adjustable from the rear of the cabinet.

Additionally, the actuator may be equipped to provide an electrical 4 – 20 mADC gas flow transmission signal for the purpose of indicating and/or recording gas flow rate at a remote location. When this option is used, both vacuum switches are provided to serve as a means of biasing the signal to zero flow condition(4 mADC) during any operating period in which the operating vacuum becomes inadequate or excessive; each condition being associated with a loss of gas flow.

第六章　工程法律法规

2021年,习近平在主持十九届中共中央政治局第三十次集体学习时强调,要深刻认识新形势下加强和改进国际传播工作的重要性和必要性,下大气力加强国际传播能力建设;习近平在主持中共中央政治局第三十五次集体学习时强调,要坚持统筹推进国内法治和涉外法治……讲好中国法治故事。

第一节　背景分析

随着"一带一路"倡议不断推进和国际工程项目的快速发展,相关法律法规等资料的翻译也成为国际工程项目中翻译工作内容的重要组成部分,翻译的相关法律文本的数量均呈几何级数增长。

法律法规的特性决定了相关翻译的特殊性。文本具有典型的强制性和严肃性,要求译者除了具备必要的翻译技能,还须具备一定的法律思维认知、法律语言和法律知识等综合能力。

国际工程涉及的法律法规或具有法律效力的合同、协议等文本,具有典型的法律语言特征和法律专业特点。译者的能力不仅仅包括翻译技能,还包括职业道德和操守。在语言服务工作中,工作内容可能涉及各类工程专业和行业背景、经济、环境、社会习俗等领域,译者应时刻谨记国家意识,保持自身立场,通晓国际法律规则,讲好中国法治故事,让世界更好地理解中国的法治方案和法治智慧,提升中国软实力。

工程法律法规翻译应当遵循法律翻译的特殊要求和原则,保持立场,措辞规范,内容精确,语言严谨。

首先,立场坚定。法律翻译者要明确立场,具备法治精神,明确语言内涵与外延,才能保证翻译公正。其次,具备专业知识。译者应当同时具备工程项目和相关法律法规知识背景,具备必要的专业知识结构,才能保证译文的正确和准确。再次,精确忠实。法律翻译中"精确"是重中之重,文字使用恰当、准确是第一要务。译者应真正做到忠实于原文,准确传递信息,避免任何语言上的歧义,以免差之毫厘、谬以千里。最后,规范严谨。译者应当具备足够的法治思维,保留法律文本的语言特色,措辞规范,语言严谨、客观,保持和原文一致的法律文体风格。

第二节　专业术语

序号	中文	英文
1	安全标准	safety standard
2	安全程序	safety procedure
3	安全规则	safety regulation
4	案件	case
5	被代理人	principal
6	被派遣的工作人员	dispatched employee
7	被侵权人	infringed person
8	变更权	right to vary
9	标的确定	the object of the contract is ascertained
10	标底	pre-tender estimate
11	剥夺公权	prescription
12	不存在侵权行为的声明	statement of non-infringement
13	不可预见的困难	unforeseeable difficulty
14	不溯既往性	nonretroactive character
15	财政法	fiscal law
16	残疾赔偿金	disability compensation
17	超越权限	act ultra vires
18	撤回要约	revocation of offer
19	撤销(判决)	repeal rescission
20	撤销要约	withdrawal of offer

续表

序号	中文	英文
21	诚信原则	principle of good faith
22	承包商调查	contractor to search
23	承继权利	succeed to the right
24	承揽人	contractor
25	承诺	acceptance
26	承诺的撤回	withdrawal of acceptance
27	承诺生效	acceptance becomes effective
28	惩罚性赔偿	punitive damage
29	初步证据	preliminary evidence
30	处罚	penalize
31	从权利	accessory right
32	从债务	accessory obligation
33	大气污染防治法	Law on the Prevention and Control of Atmospheric Pollution
34	担保	bond
35	抵消	offset
36	定标	bid selection
37	定做人	ordering party
38	罚则	penalty provision
39	法案	draft
40	法定代表人	legal representative
41	法定货币	lawful currency
42	法规	regulation
43	法令	decree
44	法律汇编	codification
45	非法人组织	unincorporated organization
46	废除（法律）	repeal, revocation, annulment
47	废除（合同）	cancellation, annulment, invalidation

续表

序号	中文	英文
48	废止/取消	abolish
49	分担损失	share losses
50	分立	split
51	分期支付	payment in installment
52	否决	veto
53	个人独资企业法	Sole Proprietorship Enterprise Law
54	工作人员	employee
55	公平交易/平等竞争	trade and compete on a fair and equal basis
56	共同责任	joint liability
57	关税	duty
58	国际法	international law
59	过错	fault
60	海洋环境保护法	Marine Environment Protection Law
61	合并	merge
62	合同	contract
63	合同内容	contract content
64	合同条款	contract clause
65	互联网信息服务管理办法	Measures for Administration of Information Service via Internet
66	会计法	Law on Accountancy
67	豁免/豁免权	immunity
68	集成电路布图设计保护条例	Regulations on the Protection of Layout-Designs of Integrated Circuits
69	技术规范	technical specification
70	价款或者报酬	price or remuneration
71	监督	supervise
72	监督管理	supervision and regulation
73	监护人	guardian

续表

序号	中文	英文
74	监护职责	duties of a guardian
75	建设单位	construction unit
76	交付期限	delivery period
77	接受劳务派遣的用工单位	employer receiving the dispatched employee
78	接受劳务一方	party receiving labor service
79	解除合同	rescind the contract
80	解除权	the right to rescission
81	解除权行使期限	the time limit for exercising the right to rescission
82	解除事由	the cause for rescission
83	解决争议	dispute resolution
84	解决争议方法	dispute resolution method
85	进场通路	access route
86	进入权	right of access
87	近亲属	close relative
88	精神损害赔偿	compensation for pains and suffering
89	开标	bid opening
90	扣押	attachment
91	扣押物	distress
92	劳动争议仲裁申请书	petition for labor dispute arbitration
93	劳工法	labor law
94	劳务派遣	labor dispatch
95	劳务派遣单位	employer dispatching the employee
96	立法	legislation
97	立法法	legislation law
98	连带债权	joint and several claim
99	连带债务	joint and several obligation
100	履行地点	place of performance
101	履行方式	manner of performance

续表

序号	中文	英文
102	履行费用	cost of performance
103	履行合同	perform the contract
104	履行期限	time period of performance
105	免除债务	exempt the obligation
106	免责条款	exculpatory clause
107	民法	civil law
108	民防工程建设和使用管理办法	Procedures on Administration of Construction and Use of Civil Defense Projects
109	民事法律关系	civil juristic relationship
110	民事关系	civil-law relation
111	民事权益	civil-law rights and interests
112	民事行为能力人	person with capacity for performing civil juristic acts
113	民事主体	person of the civil law
114	评标	bid evaluation
115	起诉	sue/litigate/prosecute
116	气象法	meteorology law
117	侵权人	tortfeasor
118	侵权行为	tortious act
119	侵权责任	tort liability
120	权利人	right holder
121	权益	interest
122	人身意义的特定物	object of personal significance
123	森林法实施条例	Regulations for the Implementation of Forestry Law
124	上市公司治理准则	Code of Corporate Governance for Listed Companies in China
125	上诉	appeal
126	身份关系	personal relationship
127	审查	examine

续表

序号	中文	英文
128	施工要求	construction requirement
129	实际履行地	place of actual performance
130	实施	conduct
131	实现债权的有关费用	the relevant expenses incurred by the creditor for enforcing his claim
132	实质性变更	material alteration
133	市场秩序	market order
134	授权	authorize
135	水利产业政策	water conservancy industrial policy
136	水污染防治法实施细则	Rules for Implementation of the Law on the Prevention and Control of Water Pollution
137	死亡赔偿金	death compensation
138	损害	undermine
139	损害赔偿	damage compensation
140	提存	place in escrow
141	提供劳务一方	party providing labor service
142	条款	clause
143	通知到达	reach the other party
144	网络服务提供者	network service provider
145	网络用户	network user
146	危险化学品安全管理条例	Regulations on the Control over Safety of Dangerous Chemicals
147	违规	violation
148	未履行的义务	unfulfilled obligations
149	无民事行为能力人	person with no capacity for performing civil juristic acts
150	无权代理人	unauthorized agent
151	无资格	legal incapacity

续表

序号	中文	英文
152	误工减少的收入	loss of income due to absence from work
153	现场清理	clearance of site
154	享有选择权	have the right of choice
155	新化学物质环境管理办法	Provisions on the Environmental Administration of New Chemical Substances
156	新要约	new offer
157	行使选择权	exercise the right of choice
158	行为人	actor
159	行政法	administrative law
160	许可	permits
161	要约邀请	invitation to offer
162	一般义务	general obligations
163	一次性支付	lump-sum payment
164	医疗费	medical expenses
165	议标	bid negotiation
166	银行条例	Banking Ordinance
167	用人单位	employer
168	逾期交付/逾期提取	overdue delivery
169	预期寿命期	intended working life
170	责任主体	subject of liability
171	债权债务同归于一人	the claim and obligation are merged to be held by the same person
172	招标	bid invitation
173	真实身份	real identity
174	执行政府定价或者政府指导价	implement the government-set or government-guided price
175	质量保证	quality assurance
176	质量要求	quality requirement

续表

序号	中文	英文
177	中华人民共和国产品质量法	Product Quality Law of the People's Republic of China
178	中华人民共和国合同法	Contract Law of the People's Republic of China
179	中华人民共和国土地管理法实施条例	Regulations on the Implementation of the Land Administration Law of the People's Republic of China
180	中华人民共和国宪法修正案	Amendment to the Constitution of the People's Republic of China
181	中华人民共和国行政复议法	Law of the People's Republic of China on Administrative Reconsideration
182	中华人民共和国职业病防治法	Code of Occupational Disease Prevention of PRC
183	中华人民共和国专利法	Patent Law of the People's Republic of China
184	仲裁裁决	award
185	仲裁庭	arbitration tribunal
186	仲裁委员会	arbitration committee
187	主债务	principal obligation
188	主张抵销	claim a set-off
189	专利法实施细则	Implementing Regulations of the Patent Law
190	追偿	right of recourse
191	资金安排	financial arrangement
192	资质	qualification
193	自动解除	automatically rescinded
194	自然法	natural law
195	总则	general provisions
196	最后裁决书	final award
197	遵守	comply with
198	作废（支票）	cancellation

第三节　翻译案例

案例一：中译英

原文：

第十八章　建设工程合同

第七百八十八条　建设工程合同是承包人进行工程建设,发包人支付价款的合同。

建设工程合同包括工程勘察、设计、施工合同。

第七百八十九条　建设工程合同应当采用书面形式。

第七百九十条　建设工程的招标投标活动,应当依照有关法律的规定公开、公平、公正进行。

第七百九十一条　发包人可以与总承包人订立建设工程合同,也可以分别与勘察人、设计人、施工人订立勘察、设计、施工承包合同。发包人不得将应当由一个承包人完成的建设工程支解成若干部分发包给数个承包人。

总承包人或者勘察、设计、施工承包人经发包人同意,可以将自己承包的部分工作交由第三人完成。第三人就其完成的工作成果与总承包人或者勘察、设计、施工承包人向发包人承担连带责任。承包人不得将其承包的全部建设工程转包给第三人或者将其承包的全部建设工程支解以后以分包的名义分别转包给第三人。

禁止承包人将工程分包给不具备相应资质条件的单位。禁止分包单位将其承包的工程再分包。建设工程主体结构的施工必须由承包人自行完成。

第七百九十二条　国家重大建设工程合同,应当按照国家规定的程序和国家批准的投资计划、可行性研究报告等文件订立。

第七百九十三条　建设工程施工合同无效,但是建设工程经验收合格的,可以参照合同关于工程价款的约定折价补偿承包人。

建设工程施工合同无效,且建设工程经验收不合格的,按照以下情形处理:

(一)修复后的建设工程经验收合格的,发包人可以请求承包人承担修复费用;

(二)修复后的建设工程经验收不合格的,承包人无权请求参照合同关于工程价款的约定折价补偿。

发包人对因建设工程不合格造成的损失有过错的,应当承担相应的责任。

第七百九十四条　勘察、设计合同的内容一般包括提交有关基础资料和概预算等文件的期限、质量要求、费用以及其他协作条件等条款。

第七百九十五条　施工合同的内容一般包括工程范围、建设工期、中间交工工程的开工和竣工时间、工程质量、工程造价、技术资料交付时间、材料和设备供应责任、拨款和结算、竣工验收、质量保修范围和质量保证期、相互协作等条款。

第七百九十六条　建设工程实行监理的,发包人应当与监理人采用书面形式订立委托监理合同。发包人与监理人的权利和义务以及法律责任,应当依照本编委托合同以及其他有关法律、行政法规的规定。

第七百九十七条　发包人在不妨碍承包人正常作业的情况下,可以随时对作业进度、质量进行检查。

第七百九十八条　隐蔽工程在隐蔽以前,承包人应当通知发包人检查。发包人没有及时检查的,承包人可以顺延工程日期,并有权请求赔偿停工、窝工等损失。

第七百九十九条　建设工程竣工后,发包人应当根据施工图纸及说明书、国家颁发的施工验收规范和质量检验标准及时进行验收。验收合格的,发包人应当按照约定支付价款,并接收该建设工程。

建设工程竣工经验收合格后,方可交付使用;未经验收或者验收不合格的,不得交付使用。

第八百条　勘察、设计的质量不符合要求或者未按照期限提交勘察、设计文件拖延工期,造成发包人损失的,勘察人、设计人应当继续完善勘察、设计,减收或者免收勘察、设计费并赔偿损失。

第八百零一条　因施工人的原因致使建设工程质量不符合约定的,发包人有权请求施工人在合理期限内无偿修理或者返工、改建。经过修理或者返工、改建后,造成逾期交付的,施工人应当承担违约责任。

第八百零二条　因承包人的原因致使建设工程在合理使用期限内造成人身损害和财产损失的,承包人应当承担赔偿责任。

第八百零三条　发包人未按照约定的时间和要求提供原材料、设备、场地、资金、技术资料的,承包人可以顺延工程日期,并有权请求赔偿停工、窝工等损失。

第八百零四条 因发包人的原因致使工程中途停建、缓建的,发包人应当采取措施弥补或者减少损失,赔偿承包人因此造成的停工、窝工、倒运、机械设备调迁、材料和构件积压等损失和实际费用。

第八百零五条 因发包人变更计划,提供的资料不准确,或者未按照期限提供必需的勘察、设计工作条件而造成勘察、设计的返工、停工或者修改设计,发包人应当按照勘察人、设计人实际消耗的工作量增付费用。

第八百零六条 承包人将建设工程转包、违法分包的,发包人可以解除合同。

发包人提供的主要建筑材料、建筑构配件和设备不符合强制性标准或者不履行协助义务,致使承包人无法施工,经催告后在合理期限内仍未履行相应义务的,承包人可以解除合同。

合同解除后,已经完成的建设工程质量合格的,发包人应当按照约定支付相应的工程价款;已经完成的建设工程质量不合格的,参照本法第七百九十三条的规定处理。

第八百零七条 发包人未按照约定支付价款的,承包人可以催告发包人在合理期限内支付价款。发包人逾期不支付的,除根据建设工程的性质不宜折价、拍卖外,承包人可以与发包人协议将该工程折价,也可以请求人民法院将该工程依法拍卖。建设工程的价款就该工程折价或者拍卖的价款优先受偿。

第八百零八条 本章没有规定的,适用承揽合同的有关规定。

（摘自《中华人民共和国民法典》第三编第二分编第十八章）

译文：

Chapter XVIII

Contracts for Construction Project

Article 788

A contract for construction project is a contract under which a contractor carries out the construction of a project and the contract-offering party pays the price in return.

Contracts for construction project consist of contracts for project prospecting, designing, and construction.

Article 789

A contract for construction project shall be made in writing.

Article 790

Bidding for a construction project shall be carried out in an open, fair, and impartial manner in accordance with the provisions of the relevant laws.

Article 791

A contract-offering party may conclude a contract for construction project with a general contractor, or conclude separate contracts for prospecting, designing, and construction with the prospecting, designing, and construction parties respectively. A contract-offering party may not break up one construction project that should be completed by one contractor into several parts and offer them to several contractors.

A general contractor or a prospecting, designing, or construction contractor may, upon consent by the contract-offering party, entrust part of the contracted work with a third person. The third person shall assume joint and several liability with the general contractor or the prospecting, designing, or construction contractor to the contract-offering party on the work product of the third person. A contractor may not delegate the whole of the contracted construction project to a third person or break up the contracted construction project into several parts and delegate them separately to third persons in the name of subcontracting.

A contractor is prohibited from subcontracting the contracted project to any entity without the corresponding qualifications. A subcontractor is prohibited from re-subcontracting the contracted project. The main structure of the construction project must be completed by the contractor itself.

Article 792

Contracts for major construction projects of the State shall be concluded in accordance with the procedures set forth by the State and such documents as investment plans and feasibility study reports approved by the State.

Article 793

Where a contract for construction project is invalid but the construction project has passed the inspection for acceptance, the contractor may be compensated, with reference to the project price agreed in the contract, based on the appraised price of the construction project.

Where a contract for construction project is invalid and the construction project fails to pass the inspection for acceptance, it shall be dealt with in accordance with the following provisions:

(1) where the construction project after being repaired has passed the inspection for acceptance, the contract-offering party may request the contractor to bear the repairing costs; or

(2) where the construction project after being repaired still fails to pass the inspection for acceptance, the contractor has no right to request for payment with reference to the project price agreed in the contract or based on the appraised price of the construction project.

Where a contract-offering party is at fault for the loss caused by the substandard of the construction project, he shall bear corresponding liabilities.

Article 794

A prospecting or designing contract generally contains clauses specifying the time limit for submission of documents relating to the basic materials and budget, quality requirements, expenses and other cooperative conditions, and the like.

Article 795

A construction contract generally contains clauses specifying the scope of the project, the period for construction, the time of commencement and completion of the project to be delivered in midcourse, project quality, costs, delivery time of technical materials, the responsibility for the supply of materials and equipment, fund allocation and settlement, project inspection and acceptance upon its completion, range and period of quality warranty, cooperation, and the like.

Article 796

For any construction project to which a superintendence system is applied, the contract-offering party shall conclude an entrustment contract of superintendence in writing with the entrusted superintendent. The rights and obligations as well as the legal liabilities of the contract-offering party and the superintendent shall be defined in accordance with the provisions on entrustment contracts of this Book as well as the relevant provisions of other laws and administrative regulations.

Article 797

The contract-offering party may, without disturbing the normal operation of the contractor, inspect the progress and quality of the work at any time.

Article 798

Prior to the concealment of a concealed project, the contractor shall notify the contract-offering party to inspect it. If the contract-offering party fails to conduct an inspection in a timely manner, the contractor may extend the period for the completion of the project accordingly, and may request compensation for the losses caused by the work stoppage, the workers' forced idleness, and the like.

Article 799

Upon completion of a construction project, the contract-offering party shall promptly undertake the inspection for acceptance in accordance with the construction drawings and descriptions, as well as the rules of inspection and acceptance of construction projects and the standards for quality inspection issued by the State. Where the project passes the inspection for acceptance, the contract-offering party shall pay the agreed price and take over the construction project.

A construction project may be delivered and put into use only after it has passed the inspection for acceptance upon completion. Without being inspected or failing to pass the inspection, the construction project may not be delivered or put into use.

Article 800

Where losses are caused to a contract-offering party due to the fact that the prospecting or designing does not conform to the quality requirements or that the prospecting or designing documents are not submitted as scheduled, so that the period for construction is delayed, the prospecting or designing party shall continue on perfecting the prospecting or designing, reduce or waive the prospecting or designing fees, and make compensation for the losses.

Article 801

Where the quality of a construction project fails to conform to the contract due to a reason attributable to the constructor, the contract-offering party has the right to request the constructor to repair, rework, or reconstruct the project without further charge within a reasonable period of time. If delivery is delayed because of the repair, reworking, reconstruction, the constructor shall bear default liability.

Article 802

Where a construction project causes personal injury and property damage within a reasonable period of use of the project due to a reason attributable to the contractor, the contractor shall bear the liability for compensation.

Article 803

Where a contract-offering party fails to provide raw materials, equipment, premises, funds, or technical materials at the agreed time and pursuant to the agreed requirements, the contractor may extend the period of construction accordingly and has the right to request compensation for the losses caused by work stoppage, workers' forced idleness, and the like.

Article 804

If a construction project is stopped or suspended in midcourse due to a reason attributable to the contract-offering party, the contract-offering party shall take measures to make up for or mitigate the loss, and compensate the contractor for any losses caused and actual expenses incurred by work stoppage, workers' forced idleness, back transportation, transfer of machinery equipment, the backlog of materials and structural components, and the like.

Article 805

Where a contract-offering party alters his plan, provides inaccurate materials, or fails to provide necessary working conditions for prospecting or designing according to the schedule, thus causing the redoing or stoppage of the prospecting or designing work or the revision of the design, the contract-offering party shall pay additional fees according to the amount of work actually undertaken by the prospecting or designing party.

Article 806

Where a contractor delegates or illegally subcontracts the construction project to others, the contract-offering party may rescind the contract.

Where the main construction materials, construction components and accessories, and equipment provided by the contract-offering party fail to conform to the mandatory standard, or the contract-offering party fails to perform the obligation of providing assistance, so that the contractor cannot undertake the construction work, if the contract-offering party still fails to perform the corresponding obligations within a reasonable period of time after being demanded, the contractor may rescind the contract.

Where, after the contract is rescinded, the quality of the completed construction project is found to be up to standard, the contract-offering party shall make corresponding payment for the construction project in accordance with the agreement. If the quality of the completed construction project is found to be substandard, the provisions of Article 793 of this Code shall be applied *mutatis mutandis*.

Article 807

Where a contract-offering party fails to pay the price in accordance with the agreement, the contractor may demand the contract-offering party to make the payment within a reasonable period of time. Where the contract-offering party still fails to pay the price upon expiration of the said period, the contractor may negotiate with the contract-offering party to appraise the construction project, or request the people's court to sell the project through auction in accordance with law, unless the construction project is by its nature unsuitable for appraisal or auction. The payment for the construction of the project shall be satisfied, in priority, from the proceeds obtained from the appraisal or auction of the said project.

Article 808

For matters not provided in this Chapter, the relevant provisions on work contracts shall be applied.

案例二：英译中

原文：

Oil Pollution Act of 1990

In March 1989, the Exxon Valdez discharged over 11 million gallons of crude oil into the pristine waters of Prince William Sound in Alaska. The dramatic television footage, and the perceived slow and inadequate response to the spill, pushed Congress to enact the Oil Pollution Act of 1990 (OPA). In truth, the OPA was the culmination of over a decade of work by Congress to consolidate oil spill response authorities under various federal laws, including section 311 of the CWA. The OPA expanded prevention and preparedness requirements for oil spills, improved response capabilities, increased the limits of liability for discharges of oil, and established expanded research and development funds related to oil spill impacts and response. The OPA also established a new Oil Spill Liability Trust Fund (OSLTF). Subtitles B and C of Title IV of the OPA extensively amended section 311. For purposes of this chapter, discussion of the OPA is limited to those provisions that directly affect section 311.

Definitions in CWA Section 311

"Discharge" is defined broadly in section 311 to include any spilling, leaking, pumping, pouring, emitting, emptying, or dumping of oil or hazardous substances.

The definition of "oil" in section 311 has been carried over from the WQIA as discussed above. This definition is not limited to petroleum products and thus encompasses oils such as animal or vegetable oils. Oil for purposes of discharge liability under the OPA is defined to exclude any kind of oil that is considered a hazardous substance subject to the CERCLA.

The EPA identifies "hazardous substances" for purposes of section 311 by specific listings in implementing regulations. To date, the EPA has designated approximately 300 chemicals as hazardous substances.

General Discharge Prohibition in CWA Section 311

Section 311 prohibits the discharge of oil or hazardous substances into or upon designated waters of the United States and adjoining shorelines in such quantities as are determined by EPA to be harmful.

The "designated waters" include navigable waters of the United States (including the territorial seas) and adjoining shorelines and the contiguous zone. The general prohibition also applies to waters beyond the contiguous zone that contain or support natural resources under the exclusive management of the United States(the EEZ) or where the discharge results from activities regulated by the Outer Continental Shelf Lands Act or the Deepwater Ports Act of 1974.

The EPA has determined that a "harmful quantity" of oil is an amount that, when discharged, violates applicable water quality standards or causes a film or sheen upon or discoloration of the surface of the water or adjoining shorelines, or causes a sludge or emulsion to be deposited beneath the surface of the water or on the adjoining shoreline.

The discharge of a hazardous substance is determined to be "harmful" if the amount meets or exceeds the designated "reportable quantity" for the substance as determined by the EPA. Reportable quantities of hazardous substances discharged range from 1 pound to 5,000 pounds, depending on factors such as the toxicity of the substance.

Exemptions from the General Discharge Prohibition of Section 311

Over the years there had been much confusion about the relationship between dis-charges requiring a National Pollutant Discharge Elimination System (NPDES) permit under section 402 of the CWA and discharges prohibited under section 311. In 1978 Congress amended section 311 to clarify this issue and to ensure that section 311 will be applied primarily to "classic spill" situations. Section 311 provides that the following discharges are not prohibited:

1. Discharges in compliance with an NPDES permit, where the discharge is subject to a pollutant-specific effluent limitation or a limitation on an indicator parameter intended to address the pollutant in question;

2. Discharges identified in the NPDES permit process, made a part of the public record, and subject to a condition in the permit;

3. Continuous or anticipated intermittent discharges from a point source, identi-

fied in a permit or permit application, which are caused by events occurring within the scope of relevant operating or treatment systems.

The latter two exemptions apply to discharges of hazardous substances regardless of whether the discharge is in compliance with a permit.

Discharges of hazardous substances from industrial facilities to publicly owned treatment works (POTWs) are not subject to regulation under section 311. However, section 311 does apply to discharges of hazardous substances to the POTW from a mobile source, such as a truck, unless the discharger has authorization from the POTW to discharge and the discharge is pretreated to comply with CWA requirements.

Discharges of oil from vessels that comply with the International Convention for the Prevention of Pollution from Ships (MARPOL 73/78) and that are into the contiguous zone of the United States or that may affect natural resources under the exclusive management of the United States are excluded from the prohibition in section 311.

Section 311 also exempts discharges in quantities and at times and locations or under circumstances and conditions deemed not to be harmful by regulation. 40 In this regard, the EPA has exempted certain discharges, including discharges of oil from properly functioning vessel engines and discharges of oil accumulated in the bilge of a vessel that comply with the requirements of MARPOL 73/78.

Spill Notification Requirement

The person in charge of the vessel or facility discharging harmful quantities of oil or a hazardous substance in violation of the discharge prohibition in section 311 must report the discharge to the appropriate federal agency as soon as that person has knowledge of the discharge. The United States has established a National Response Center (NRC), operated by the Coast Guard, to receive such reports and determine the response action (if any) that is necessary. In turn, the NRC is required to notify the affected state(s). Failure to report is subject to felony criminal sanctions. Reports filed with the NRC cannot be used against any natural person making a report in any criminal case, except in a prosecution for giving false statements or for perjury.

Spill Prevention

Section 311 mandates that the president issue regulations establishing procedures, methods, equipment, and other requirements to prevent discharges of oil and hazardous substances from vessels and facilities and to contain such discharges. By executive order, the president has delegated authority to the EPA to regulate non-transportation-related onshore facilities and has delegated to the Department of Transportation the authority to regulate vessels and marine-transportation-related facilities.

（摘自《清洁水法》）

译文：

1990 年的《油污染法》

1989 年 3 月,埃克森石油公司的瓦尔迪兹号油轮在阿拉斯加附近的威廉王子海峡触礁,致使 1100 万加仑的原油泄漏到该海域。媒体连篇累牍的报道加之公众对该事件的超常规反应迫使国会在 1990 年颁布了《油污染法》。事实上,《油污染法》的颁布是国会数十年来在不同的联邦法律下试图巩固石油泄漏应对权力的顶峰,这其中就包括《清洁水法》第 311 条。《油污染法》扩展了对石油泄漏预防和应对方案的要求,提高了应对能力,增加了对石油排放的责任限制,并成立了石油溢漫责任信托基金(OSLTF)。《油污染法》中 IV 栏目下的 B 类和 C 类对第 311 条做出了大范围的修正。本章针对 OPA 的讨论主要和直接影响第 311 条的相关条例有关。

《清洁水法》第 311 条中的定义

此章节中,泄漏在广义上包含各种形式的油脂及有害物质的溢出、泄漏、倾倒、散发、排空及倾泻等。

上文提到过第 311 条中油脂的定义照搬自《水质改善法》。该定义并不局限于石油相关产品,还包括动植物油。《油污染法》(OPA)规定排放责任所涉及油类不包含《综合环境责任赔偿和义务法》(CERCLA)所认定的有害油类。

环保局在实施条例中列出了有害物质明细。到目前为止,EPA 中已经列出了近 300 种有害的化学物质。

一般性排放禁令

第 311 条禁止油脂及有害物质在美国适航水域及海岸附近的超量排放,这些物质在环保局被界定为有害物质。

其中的指定水域包括美国的可航水域(含领海)和邻近水域。一般性禁令还适用于美国专属经济区管理下的包含或支撑自然资源存在的超邻近水域,或者是在《大陆架外围缘地法案》及1974年《深水港法案》管理下的活动所引起的泄漏结果影响范围。

环保局对于"危害数量"的量的规定是指泄漏发生后,导致水质不达标、水面或邻近水域产生覆盖薄膜及油污,或者是达到需要在水面或水域下方使用污泥和乳化剂的泄漏数量。

如果环保局列举的那些物质达到或超过规定的泄漏量,那么这些物质就被认为是具有危害性的有害物质。有害物质泄漏的需报告量从1磅到5000磅不等,这主要依据其毒性等因素来判定。

第311条中一般性禁止泄漏的豁免

长期以来,关于《清洁水法》第402条指导下建立的全国污染物质排放清除系统(NPDES)和第311条所禁止的污染物排放的关系一直存在争议。1978年,国会对第311条进行了修改,解决并明确了第311条首先适用于传统泄漏的情况。以下排放物质不在第311条所规定的禁止排放之列:

1.全国污染物质排放清除系统允许的排放物质必须遵守特殊污染物排放限量,或者控制在一定的参数之内以便进行污染处理;

2.全国污染物质排放清除系统管理下有加工处理程序、建立公共记录并符合特殊情况的排放物;

3.在相关操作或处理系统范围内发生的并得到许可的连续或可预测的阶段性泄漏。

后两项同样适用于有害物质的排放,不论其是否符合这些规定。

从工业设施流入公共污水处理系统的有害物质排放不受第311条的约束。但是,移动污染源产生的并波及公共污水处理系统(POTWs)的有害物质排放则适用第311条,除非排放者得到了公共污水处理系统的授权或者对排放物进行预处理以符合《清洁水法》的要求。

符合《防止船舶污染国际公约》(MARPOL 73/78)规定的船舶油脂泄漏和影响美国专属经济区或邻近水域自然资源的泄漏也不在第311条的禁止之列。

第311条同样豁免了特定时间、地点、环境及数量下的无害排放物。有些排放物同样也不在环保局管理范围内,这其中就包括运转正常的船舶发动机所产生的油料泄漏和符合 MARPOL73/78 要求的船底积油的排放。

溢油通知要求

违反第 311 条规定的排污体所有者或经营者必须在发现泄漏的第一时间通知相关联邦机构。美国政府成立了由海岸警卫队管理的国家快速反应中心(NRC)来接收相关报告并采取必要的应急行动。反过来,快速反应中心也要知会受影响的各州。迟报、瞒报、误报将面临重罪刑事制裁。交由快速反应中心的报告不能用于针对任何自然人的刑事诉讼中,但是错误陈述和做伪证的情况例外。

外溢预防

第 311 条授权总统颁布相关法令,通过建立相关处理程序、实施方法及设备以达到预防载有危险物质的船舶及设施物质外溢的作用。总统通过行政命令授权环保局相关职权管理非船运相关的岸上设施,船舶和船运相关设施的管理则交由交通部。

第四节　翻译评析

工程法律法规翻译注重语言的严谨和内容的客观,多以直译为主,在词汇方面多涉及工程和法律类专业术语,在句式方面多用长句,逻辑严密,在篇章方面则遵循法规条款内容要求,通常依据内容采用总分结构,保持统一法律文本的严谨。

一、词法层面

以下案例中涉及多个专业术语,术语的翻译相对固定,通常首选平行文本,寻找译文语言中相对应的术语进行替换。如无平行文本,则通常保持原文词性不变,以直译为主,多个名词并列的情况较多见。

【例1】原文:第七百九十五条　施工合同的内容一般包括工程范围、建设工期、中间交工工程的开工和竣工时间、工程质量、工程造价、技术资料交付时间、材料和设备供应责任、拨款和结算、竣工验收、质量保修范围和质量保证期、相互协作等条款。

译文：

Article 795

A construction contract generally contains clauses specifying the project scope, the period for construction, the time of commencement and completion of the project to be delivered in midcourse, project quality, costs, delivery time of technical materials, the responsibility for the supply of materials and equipment, fund allocation and settlement, project inspection and acceptance upon its completion, range and period of quality warranty, cooperation, and the like.

分析：本案例中涉及"施工合同""中间交工工程""开工""工程造价"等多个专业术语，在翻译时直接将它们替换为译语中相对应的平行文本"construction contract""the project to be delivered in midcourse""commencement""project costs"。

【例2】原文："Discharge" is defined broadly in section 311 to include any spilling, leaking, pumping, pouring, emitting, emptying, or dumping of oil or hazardous substances.

译文：此章节中，泄漏在广义上包含各种形式的油脂及有害物质的溢出、泄漏、倾倒、散发、排空及倾泻等。

分析：该案例中，原文出现"spilling""leaking""pumping""pouring""emitting""emptying""dumping"一系列动名词。这类名词术语，常见于日常生活用语。此类词语的使用环境或上下文，决定了其具有特定的行业意义，分别翻译为"溢出""泄漏""倾倒""散发""排空"和"倾泻"。

【例3】原文：The definition of "oil" in section 311 has been carried over from the WQIA as discussed above. This definition is not limited to petroleum products and thus encompasses oils such as animal or vegetable oils. Oil for purposes of discharge liability under the OPA is defined to exclude any kind of oil that is considered a hazardous substance subject to the CERCLA.

译文：上文提到过第311条中油脂的定义照搬自《水质改善法》。该定义并不局限于石油相关产品，还包括动植物油。《油污染法》（OPA）规定排放责任所涉及油类不包含《综合环境责任赔偿和义务法》（CERCLA）所认定的有害油类。

分析:该案例中出现了缩略语"WQIA""OPA"和"CERCLA",翻译时通常须将其还原为全称,分别是《水质改善法》《油污染法》和《综合环境责任赔偿和义务法》。

值得一提的是,缩略语第一次出现时通常为全称和缩略语同时出现。为方便下文行文,下文将不再出现全称,而用缩略语。在翻译时,为便于读者接受,第一次出现缩略语时也应同时写出全称和缩略语,下文重复出现的缩略语则可以译文缩略或零翻译两种形式出现。

二、句法层面

法律文本多涉及强制性内容,通常避免使用模糊性较强的措辞,如"可能""或许",而多用"应""须"等。在英文中,法律文本同样多采用预期强烈的词语,如"shall"等。

【例4】原文:建设工程竣工后,发包人应当根据施工图纸及说明书、国家颁发的施工验收规范和质量检验标准及时进行验收。验收合格的,发包人应当按照约定支付价款,并接收该建设工程。

译文:Upon completion of a construction project, the contract-offering party shall promptly undertake the inspection for acceptance in accordance with the construction drawings and descriptions, as well as the rules of inspection and acceptance of construction projects and the standards for quality inspection issued by the State. Where the project passes the inspection for acceptance, the contract-offering party shall pay the agreed price and take over the construction project.

分析:该案例中,原文措辞具有典型的"强制性"特点,多处使用"应当"一词,在英文译文中,则相应采用"shall"这一语义明确的词语。

【例5】原文:建设工程竣工经验收合格后,方可交付使用;未经验收或者验收不合格的,不得交付使用。

译文:A construction project may be delivered and put into use only after it has passed the inspection for acceptance upon completion. Without being inspected or failing to pass the inspection, the construction project may not be delivered or put into use.

分析:与上一个案例不同,该案例的原文语义内涵不具有唯一性。"建设工程竣工经验收合格"是交付使用的必需条件,并非必然导致"交付使用"的实现。因此,原文使用"可",而非"须"或"应",译文也相应翻译为"may"这一具有可能性的词语。

英语多长句,中文多短句。为符合语言表达习惯,英译中时通常采用拆分句的翻译技巧,而中译英时则多用合句这一翻译技巧。但在法律文本中,中文也同样多用长句。这一特殊的语言特征决定了英译中时,工程法律法规的翻译未必需要拆分句。

【例6】原文:The EPA has determined that a "harmful quantity" of oil is an amount that, when discharged, violates applicable water quality standards or causes a film or sheen upon or discoloration of the surface of the water or adjoining shorelines, or causes a sludge or emulsion to be deposited beneath the surface of the water or on the adjoining shoreline.

译文:环保局对于"危害数量"的量的规定是指泄漏发生后,导致水质不达标、水面或邻近水域产生覆盖薄膜及油污,或者是达到需要在水面或水域下方使用污泥和乳化剂的泄漏数量。

分析:在该案例中,原文包含多个小句,但实际上是一个句子。英译中时,长句的处理方式多依据句内成分的逻辑关系,将原文中的一个长句拆分为两个或两个以上的短句。但在中文法律文本中,如果在一个完整的句子中能够处理所有信息,则不对原句做拆分处理。在翻译过程中,长句更为正式和严肃,译文应当保持同样的文本风格。

三、篇章层面

法律法规条款通常始于总则,以事务细节条款结束。条款细则之间逻辑关系紧密,整体措辞风格严谨客观。

【例7】原文:

第七百九十三条　建设工程施工合同无效,但是建设工程经验收合格的,可以参照合同关于工程价款的约定折价补偿承包人。

建设工程施工合同无效,且建设工程经验收不合格的,按照以下情形处理:

(一)修复后的建设工程经验收合格的,发包人可以请求承包人承担修复费用;

（二）修复后的建设工程经验收不合格的，承包人无权请求参照合同关于工程价款的约定折价补偿。

发包人对因建设工程不合格造成的损失有过错的，应当承担相应的责任。

译文：

Article 793

Where a contract for construction project is invalid but the construction project has passed the inspection for acceptance, the contractor may be compensated with reference to the project price agreed in the contract and based on the appraised price of the construction project.

Where a contract for construction project is invalid and the construction project fails to pass the inspection for acceptance, it shall be dealt with in accordance with the following provisions:

（1）where the construction project after being repaired has passed the inspection for acceptance, the contract-offering party may request the contractor to bear the repairing costs;

（2）where the construction project after being repaired still fails to pass the inspection for acceptance, the contractor has no right to request for payment with reference to the project price agreed in the contract or based on the appraised price of the construction project.

Where a contract-offering party is at fault for the loss caused by the substandard of the construction project, he shall bear corresponding liabilities.

分析：在该案例中，一则条款包含五种情况——"建设工程施工合同无效，但是建设工程经验收合格的""建设工程施工合同无效，且建设工程经验收不合格的""修复后的建设工程经验收合格的""修复后的建设工程经验收不合格的"和"发包人对因建设工程不合格造成的损失有过错的"。译文中，每一种情况都用同样的引导词"where"引出，整体结构清晰，句式工整对称，逻辑关系清晰明了，风格与原文保持统一。

第五节　拓展延伸

一、中译英

第六十四条　违反本法规定,未取得施工许可证或者开工报告未经批准擅自施工的,责令改正,对不符合开工条件的责令停止施工,可以处以罚款。

第六十五条　发包单位将工程发包给不具有相应资质条件的承包单位的,或者违反本法规定将建筑工程肢解发包的,责令改正,处以罚款。

超越本单位资质等级承揽工程的,责令停止违法行为,处以罚款,可以责令停业整顿,降低资质等级;情节严重的,吊销资质证书;有违法所得的,予以没收。

未取得资质证书承揽工程的,予以取缔,并处罚款;有违法所得的,予以没收。

以欺骗手段取得资质证书的,吊销资质证书,处以罚款;构成犯罪的,依法追究刑事责任。

第六十六条　建筑施工企业转让、出借资质证书或者以其他方式允许他人以本企业的名义承揽工程的,责令改正,没收违法所得,并处罚款,可以责令停业整顿,降低资质等级;情节严重的,吊销资质证书。对因该项承揽工程不符合规定的质量标准造成的损失,建筑施工企业与使用本企业名义的单位或者个人承担连带赔偿责任。

第六十七条　承包单位将承包的工程转包的,或者违反本法规定进行分包的,责令改正,没收违法所得,并处罚款,可以责令停业整顿,降低资质等级;情节严重的,吊销资质证书。

承包单位有前款规定的违法行为的,对因转包工程或者违法分包的工程不符合规定的质量标准造成的损失,与接受转包或者分包的单位承担连带赔偿责任。

第六十八条　在工程发包与承包中索贿、受贿、行贿,构成犯罪的,依法追究刑事责任;不构成犯罪的,分别处以罚款,没收贿赂的财物,对直接负责的主管人员和其他直接责任人员给予处分。

对在工程承包中行贿的承包单位,除依照前款规定处罚外,可以责令停业整顿,降低资质等级或者吊销资质证书。

第六十九条　工程监理单位与建设单位或者建筑施工企业串通,弄虚作假、降低工程质量的,责令改正,处以罚款,降低资质等级或者吊销资质证书;有违法所得的,予以没收;造成损失的,承担连带赔偿责任;构成犯罪的,依法追究刑事责任。

工程监理单位转让监理业务的,责令改正,没收违法所得,可以责令停业整顿,降低资质等级;情节严重的,吊销资质证书。

第七十条　违反本法规定,涉及建筑主体或者承重结构变动的装修工程擅自施工的,责令改正,处以罚款;造成损失的,承担赔偿责任;构成犯罪的,依法追究刑事责任。

第七十一条　建筑施工企业违反本法规定,对建筑安全事故隐患不采取措施予以消除的,责令改正,可以处以罚款;情节严重的,责令停业整顿,降低资质等级或者吊销资质证书;构成犯罪的,依法追究刑事责任。

建筑施工企业的管理人员违章指挥、强令职工冒险作业,因而发生重大伤亡事故或者造成其他严重后果的,依法追究刑事责任。

第七十二条　建设单位违反本法规定,要求建筑设计单位或者建筑施工企业违反建筑工程质量、安全标准,降低工程质量的,责令改正,可以处以罚款;构成犯罪的,依法追究刑事责任。

第七十三条　建筑设计单位不按照建筑工程质量、安全标准进行设计的,责令改正,处以罚款;造成工程质量事故的,责令停业整顿,降低资质等级或者吊销资质证书,没收违法所得,并处罚款;造成损失的,承担赔偿责任;构成犯罪的,依法追究刑事责任。

二、英译中

Introduction and Scope

Section 402 of the Clean Water Act(CWA) regulates discharges of pollutants that are anticipated and relatively routine, and thus appropriately addressed through a permit program. Section 311 of the CWA (as amended by the Oil Pollution Act of 1990) governs prevention of, and response to, accidental releases and spills of oil and hazardous substances to waters of the United States. Section 311 complements section 402 of the CWA. Further, the provisions of section 311 address discharges resulting from accidental releases and spills of a limited category of pollutants: oil and hazardous substances.

Section 311 establishes spill prevention requirements, spill reporting obligations, and spill response planning requirements. Regulatory implementation and enforcement of section 311 are primarily the dual responsibility of the U. S. Coast Guard and the Environmental Protection Agency (EPA). The Coast Guard is primarily responsible for regulation and enforcement related to vessels and marine-transportation-related facilities, as well as the coastal zone of the United States, while the EPA is responsible for nontransportation-related facilities and the inland zone of the United States.

Historical Background

The first federal statute that specifically targeted pollution of coastal waters by oil was the Oil Pollution Act of 1924, which defined oil to include "oil of any kind or in any form including fuel oil, oil sludge, and oil refuse. "

The authority of the Oil Pollution Act of 1924 essentially remained in effect until several major oil spills from vessels in the late 1960s, and an oil spill from an oil production platform off the coast of Santa Barbara, California, in January 1969, led to the enactment of the Water Quality Improvement Act of 1970(WQIA). The WQIA was the first comprehensive effort to deal with discharges of oil and hazardous substances into the waters of the United States and was the forerunner of the current section 311.

第七章　工程学术论著

所谓学术是指较为专门的、有系统性的学问。学术论著是学术成果的载体,其内容是作者在某一科学领域中对某一课题进行潜心研究而获得的结果,具有系统性和专业性。工程类学术论著是某一工程领域的相关学术研究成果或科学研究记录或科学总结,在内容上具有较强的水利、电力等工程专业背景,理论性强。其语言区别于其他文本,是作者为完成特定学术研究任务而采用的特定学术语言。

第一节　背景分析

与其他学术作品一样,工程类学术论著具有学术性、理论性、科学性、系统性和专业性。

所谓学术性,指的是论著的学术研究内涵和价值,是作者针对某一学术研究领域中的问题进行实验、研究、探索或实证等得出的研究观点或结果;所谓理论性,指的是学术论著蕴含理论价值和意义,往往是对某一学术研究问题的理论分析,从而指导科学研究实践;所谓科学性,指的是学术研究论著研究方法科学、正确,尊重客观规律,分析问题态度严谨,实事求是;所谓系统性,指的是学术研究论著写作中,文章组织结构符合逻辑,内容符合相关理论或实证研究的内在逻辑要求;所谓专业性,指的是学术研究论著的专业和行业背景,其较强的专业性往往限定了读者群和阅读范围。

学术语言的特点体现了学术研究论著在特定的学术语境下完成交际任务的要求,文本表达的信息复杂,涉及的概念较抽象,专业理论性强,所使用的语言相应地更复杂。从交际角度来看,学术语言之所以难,是因为信息复杂,专业背景具有排他性,学术知识难度分层。因内容的客观性,学术语言通常客观、严谨、规范、精练。

工程类学术论著翻译应依据学术论著的文本特征,做到以下几点:翻译专业——词义专业、句式专业和表述专业;翻译精准——精准移植原文的信息;翻译美——保持原文学术语言风格,语言简洁、严谨、客观,结构整齐,措辞规范,逻辑严密。

第二节　专业术语

序号	中文	英文	备注
1	案例研究	case studies	
2	保护生理学	physiology of conservation	
3	暴雨洪涝灾害	rainstorm flood-waterlogging disaster	
4	本体论关系	ontological relations	
5	变形模量	deformation modulus	
6	变异系数法	coefficient of variation method	
7	辩证法	dialectic	
8	标准差椭圆	standard deviation ellipse	
9	博弈论	game theory	
10	不均匀性	heterogeneity	
11	部门效应	sector effect	
12	层级理论	hierarchy theory	
13	超个体	superorganism	
14	超生物体	superorganism	
15	冲击回波法	impact echo method	
16	储能模式	energy storage mode	
17	达尔文理论	Darwinian theory	
18	弹脆塑性模型	elastic-brittle-plastic model	
19	弹性参数	elastic parameters	
20	地理生态学	geoecology	
21	地貌植物地理学	physiognomic plant geography	

续表

序号	中文	英文	备注
22	定律	laws	
23	定律性	lawlikeness	
24	定位	orientation	
25	动物行为学	ethology	
26	多维的	multi-dimensional	
27	多维超空间	n-dimensional hyperspace	
28	多要素比较分析	multi-factor comparative analysis	
29	多元论	pluralism	
30	多主体应对措施	multi-agency countermeasures	
31	繁殖	reproduction	
32	反机械论的	anti-mechanistic	
33	反距离加权法	inverse distance weighting method	
34	反馈机制	feedback mechanism	
35	分离性论文	separability thesis	
36	复变函数解	complex function solution	
37	赋权试验	weighting test	
38	干扰机制	disturbance regime	
39	纲目分类法	categorical classification	
40	格式塔	gestalt	
41	个体生态学	autecology	
42	共同体	communities	
43	共享储能	shared energy storage	
44	建构主义的	constructivist	
45	关系生理学	physiology of relations	
46	观测研究	observational studies	
47	广义塑性位势理论	generalized plastic potential theory	
48	归纳法	generalization	
49	还原论	reductionism	

续表

序号	中文	英文	备注
50	洪水激增规律	flood surge law	
51	环境场	environmental field	
52	环境伦理学	environmental ethics	
53	生态学	bionomics	
54	基本概念	basic conceptions	
55	基要派	fundamentalists	
56	激光点云测量技术	laser point cloud measurement technology	
57	极端气温事件	extreme temperature events	
58	集合论	set theory	
59	集合体	assemblages	
60	技术效应	technology effect	
61	加卸载准则	loading and unloading criterion	
62	剪切模量	shear modulus	
63	剪切屈服函数	shear yield function	
64	经济效应	economic effect	
65	经验反驳	empirical refutations	
66	竞争排斥原则	competitive exclusion principle	
67	巨灾保险体制	catastrophe insurance system	
68	科学哲学	philosophy of science	
69	客观主义	objectivism	
70	空间自相关法	spatial autocorrelation method	
71	控制论者	cybernetics group	
72	控制特性	controlling nature	
73	昆虫学	entomology	
74	浪漫特性	romantic nature	
75	浪漫整体型	romantic-holistic type	
76	离域性	delocalized nature	
77	鲁宾涅特解	Rubin's solution	

续表

序号	中文	英文	备注
78	模型导向式的	model-driven	
79	牡蛎滩	oyster bank	
80	能量	energy	
81	泥沙防治	prevention of sediment	
82	偶然性	contingency	
83	耦合	coupling	
84	驱动经济	an economy of drives	
85	驱动特性	drive nature	
86	全过程管理	whole-process management	
87	缺额负荷	deficit load	
88	群落	societies	
89	人类中心主义者	anthropocentrists	
90	人员疏散路径	personnel evacuation route	
91	容量分配机制	capacity allocation mechanism	
92	三元模式	triadic pattern	
93	熵减少	entropic decline	
94	熵权法	entropy weight method	
95	摄动解	perturbation solution	
96	生境	niche	
97	生命悖论	life paradox	
98	生态定量理论	quantitative theories of ecology	
99	生态多元化	ecological plurality	
100	生态公民权利	ecological citizenship	
101	生态圈	ecosphere	
102	生态素养	ecological literacy	
103	生态系统多样性	ecosystem diversity	
104	生态学	oecologie	
105	生态知识	ecological knowledge	

续表

序号	中文	英文	备注
106	生态中心主义者	ecocentrists	
107	生物地理学	chorology	
108	生物多样性	biodiversity	
109	生物实体	biological entities	
110	湿润比	percentage of wetted area	
111	湿周	wetted perimeter	
112	十字板剪切试验	vane shear test	
113	时效损伤函数	time-dependent damage function	
114	实物工程量	real work quantity	
115	示踪模型	tracer model	
116	势波	potential wave	
117	势流	potential flow	
118	势能	potential energy	
119	势涡	potential vortex	
120	事故备用容量	reserve capacity for accident	
121	试验处理	treatment of experiment	
122	试验端子	test terminal	
123	试验项目	testing item	
124	试验小区	experimental block	
125	试运行	test run	
126	视准线法	collimation line method	
127	收敛约束法	convergence-confinement method	
128	输入功率试验	input test	
129	输沙量	sediment runoff	
130	输沙率	sediment discharge	
131	数据导向	data-driven	
132	数字地面模型	digital terrain model（DTM）	
133	甩负荷试验	load-rejection test	

续表

序号	中文	英文	备注
134	双屈服面本构模型	double yield surface constitutive model	
135	水泵参数与特性	parameters and characteristics of pump	
136	水动力学	hydrodynamics	
137	水斗式水轮机（培尔顿式水轮机）	pelton turbine	
138	水工建筑物	hydraulic structure	
139	水灰比	water-cement ratio	
140	水静力学	hydrostatics	
141	水库测量	reservoir survey	
142	水库调度	reservoir operation	
143	水库年限	ultimate life of reservoir	
144	水库水文测验	reservoir hydrometry	
145	水库塌岸	bank ruin of reservoir	
146	水库淹没实物指标	material index of reservoir inundation	
147	水库淤积测量	reservoir accretion survey	
148	太沙基渗透公式	Terzaghi permeability formula	
149	碳循环	carbon cycle	
150	体积屈服函数	volume yield function	
151	统一强度理论	unified strength theory	
152	投机猜测	speculative conjectures	
153	突变论	catastrophe theory	
154	图式化	schematization	
155	外部生理学	external physiology	
156	微电网	micro-grid	
157	微观湖泊	microcosm lake	
158	微观世界	microcosm	
159	问题机	question machines	
160	物业开发层	property development floor	

续表

序号	中文	英文	备注
161	物种多样性	species diversity	
162	物种迁移	species' migrations	
163	细胞学	cytology	
164	先验背景	transcendental context	
165	现象学概念	phenomenological concept	
166	线性倾向估计法	linear tendency estimation method	
167	相似度权重	similarity weight	
168	相似离度	similarity deviation	
169	信息间隙决策理论	information gap decision-making theory（IGDT）	
170	行动理论	action theory	
171	形态生物学	morphology	
172	研究纲领	research programmes	
173	央地协作管理	central-local collaboration management	
174	一体化管理	integrated management	
175	一元论	monism	
176	遗传多样性	genetic diversity	
177	因果关系的	causal	
178	因果循环机制	circular causal systems	
179	硬能源路径	hard energy paths	
180	庸俗唯物主义变化	vulgar materialistic variation	
181	有限差分法	finite difference method	
182	元叙事	meta-narrative	
183	运筹学	operations research	
184	真实实验	real-world experiments	
185	振动	oscillation	
186	整合层次理论	theory of integrative levels	
187	整合概念	integrative concepts	
188	植物区系地理学	floristic plant geography	

续表

序号	中文	英文	备注
189	植物生理学	plant physiology	
190	植物生态学	plant ecology	
191	装机容量	installed capacity	
192	缀块	patch	
193	准实验研究	quasi-experimental studies	
194	资源禀赋效应	resource endowment effect	
195	子域	sub-fields	
196	自然经济	economy of nature	
197	自然平衡	balance of nature	
198	自然特征	natural character	
199	自然滞洪	free flood retarding	
200	最优相似系数	optimal similarity coefficient	

第三节　翻译案例

案例一

原文：

降雨型滑坡的机理及其启示

　　摘　要：按照有效应力原理，降雨引发的规模较小的浅层滑坡，主要受岩土体的内聚力控制，而规模较大的深层滑坡则主要受摩擦力控制。对于大多数的降雨型滑坡易发地区，就滑坡预报而言，预报日前期的降雨量和土体含水量观测是非常重要的。影响滑坡的临界孔隙压力的因素在空间上是变化的。因此，在降雨条件相同时，在不同地点降雨引发滑坡的概率不同。根据目前物理学对颗粒物质的认识，可以推测：降雨引发的斜坡破坏是复杂的非线性系统。如果把降雨型滑坡的发生过程看作一个系统，降雨和环境因素作为它的输入，滑坡发生概率作为其输出，则输入与输出之间的关系是非线性的。由此，可以认为在降雨型滑坡预报中采用非线性的方法将比确定性的模型或一般线性统计方

关键词:降雨型滑坡;有效应力;颗粒流;非线性系统

译文:

Mechanism of Rainfall-Induced Landslides and its Implications to Landslide Prediction

Abstract: According to the principle of effective stress, for rainfall-induced landslides, the small and shallow landslides are mostly controlled by the cohesive strength; the large and deep landslides are controlled by friction. For the prediction of rainfall-triggered landslides, the measuring of antecedent precipitation and soil moisture is a highly important issue for most landslide-prone areas. Because under a given rainfall condition, the various environmental factors affecting critical pore-water pressure needed to trigger landslides, such as geology, topography, soil deposits, land cover and so on, vary spatially in a region, the probability of rainfall-induced landslide occurrence varies from location to location. From the understanding of granular flow in current physics, it is suggested that landslides induced by the rainfall are involved in a complex non-linear system. If rainfall and the environmental factors are taken as input to the system, landslides are considered as the output variables of the system, and then the relationship between the input and the output is non-linear. Thus, the forecast issues for rainfall-induced landslides should be treated with a non-linear method rather than traditional statistical procedures and process-based deterministic models.

Keywords: rainfall-triggered landslides; effective stress; granular flow; non-linear system

案例二

原文:

降雨型滑坡预报新方法

摘　要:详细研究了三峡地区部分县市的滑坡和降雨历史资料,从滑坡与降雨量、暴雨以及降雨时间 3 个不同角度的关系分析了降雨与降雨型滑坡的关系。在此基础上,提出了降雨因子的概念。同时,还提出了一种预报降雨型滑坡的新方法,定量化地描述了降雨型滑坡的易发程度。按照一定的标准,对每种降雨分因子进行分级,通过多因子叠合分析来研究降雨因子与降雨型滑坡之间的关系,并据此准确地预报滑坡的易发程度。通过将这种滑坡预报新方法应用于三峡的万县地区,证明了它可以比较准确地确定滑坡发生的时间。这种滑坡预报方法将为根据历史降雨和滑坡资料来预测降雨型滑坡奠定良好基础。

关键词:工程地质;降雨型滑坡;降雨因子;叠合;预报

译文:

New Method of Predicting of Rainfall-Induced Landslides

Abstract: Historical data of landslides and rainfall of some counties and cities in the Three Gorges area are investigated in detail, and the relationship between rainfall and rainfall-induced landslides is analyzed from three different points of view, rainfall, rainstorm and rainfall duration. The conception of rainfall-factor is suggested thereafter. At the same time, a new method of predicting rainfall-induced landslides is proposed to depict the probability of rainfall-induced landslides quantitatively. Every rainfall-sub-factor is graded according to some criteria, and then the relationship between rainfall-factor and rainfall-induced landslides is studied. It is proved that this new method of landslide prediction can predict the occurrence time of landslides with higher precision through applying it to Wanxian, a city in the Three Gorges area. This method of landslides prediction would lay a solid foundation for predicting rainfall-induced landslide according to historical data of rainfall and landslides.

Key words: engineering geology; rainfall-induced-landslides; rainfall-factor; superposition; prediction

案例三

原文:

基于 MIKE 21 的圩区河网水动力调控方法研究

摘　要:【目的】我国南方平原地区地势低缓,河网结构复杂。为快速解决该地区因河网水动力不足而引起的水环境问题,急需研究河网水体运行状况调控方案。【方法】以太湖流域南浔区沈庄漾"溇港圩田"河网作为研究对象,构建 Mike 21 FM 二维河网水动力模型,结合研究区河网改造后的水利工程布置和调度规则,提出定位增压的 7 种水动力调控方案,模拟分析河网在不同方案下的流场分布规律,探究圩区河网水动力提升和流场改善的优化调度方法。【结果】模拟结果表明:活水和引排水方案虽能有效改善河网水动力,但对于中心水塘的水动力驱动效果不佳。在开启西南侧河道水泵和西部泵闸的情况下,河网内大部分区域的流速基本能维持在 0.05 m/s 左右,部分河道的流速能达到 0.3 m/s 以上,并且对中心水塘水体驱动效果最好。【结论】综合分析表明,所设计的 7 种调控方案均能有效地提升河网水动力,在开启西南侧河道的 1 台水泵和西部泵闸时,流场分布最均匀且水动力整体提升效果最佳。研究结果可为南方平原河网的水利工程水动力调控提供参考。

关键词:水动力模型;调控方案;圩区河网;流场分布

译文:

Research on the Hydrodynamic Regulation Method of River Network in Polder area Based on MIKE 21

Abstract: [Objective] The southern plain area of China is low and gentle, and the river network structure is complex. In order to quickly solve the water environment problems caused by the lack of water power in the river network in this area, it is urgent to study the regulation scheme of the river network water operation. [Methods] This study took the "Lougang Polder" river network in Shenzhuangyang in Nanxun District of the Taihu Lake Basin as the research object, constructed the hydrodynamic model of Mike 21 FM two-dimensional river network, combined with the layout and scheduling rules of water conservancy projects after the transformation of the river network in the study area, proposed seven hydrodynamic regulation schemes for positioning and pressurization, simulated and analyzed the distribution of flow fields under different schemes of river net-work, and explored the optimal scheduling method of hydrodynamic improvement and flow field improvement of river network in the polder area. [Results] The simulation results show that although the live water and water diversion and drainage scheme can effectively improve the hydrodynamic force of the river network, the hydrodynamic driving effect of the central pond is not good. With the southwest river water pump and the west pump gate open, the flow velocity in most areas of the river network can basically maintain around 0.05 m/s, and the flow velocity in some rivers can reach more than 0.3 m/s, and the driving effect on the central pond water body is the best. [Conclusion] The comprehensive analysis shows that the seven control schemes designed can effectively improve the hydrodynamic force of the river network. When one water pump and the west pump gate are opened in the southwest river channel, the flow field distribution is the most uniform and the overall hydrodynamic force improvement effect is the best. The research result can provide a reference for the hydrodynamic regulation of water conservancy projects in the southern plain river network.

Keywords: hydrodynamic model; regulation scheme; polder river network; flow field distribution

案例四
原文：

东江流域下游水文情势的改变度研究

摘　要：【目的】水利工程的修建和运行会改变流域下游的天然水文过程，造成生态影响。定量评价水利工程对流域水文过程的影响，是完善工程调度和恢复下游生态流量过程的基础。针对目前流域水文改变度研究中对水库调度的影响分析不充分的问题，提出一种新的研究思路。【方法】以广东东江流域为例，采用耦合水库调度过程的分布式水文模型，以现行水库调度方式下的流量为基础，计算并分析水库修建后和水库不同下泄流量情境下东江下游博罗站的水文过程。评估在相同气象和下垫面条件下水库的修建运行对流域下游水文情势的影响程度。【结果】结果显示：(1)东江流域新丰江、枫树坝和白盆珠三大水库的修建和运行显著改变了下游博罗站的月流量过程、年极值流量以及年内径流改变率等水文指标，尤其是高低流量出现频率、流量增加率减少率和流量逆转次数改变较大；(2)新丰江单库运行阶段的整体水文改变度为70.90%，枫树坝和新丰江双库调度阶段的整体改变度为70.96%，三大水库联合调度阶段的整体水文改变度为71.77%，均属于高度改变；(3)提高上游水库的月最小下泄流量阈值,下游的水文改变度较为明显,其中新丰江坝下水文改变程度最显著。【结论】结果表明:分布式水文模型耦合水库调度和水文改变指标变化范围法,是评价流域水文情势变化的重要手段,对于缺少水库修建前实测径流数据的流域水文情势研究具有较大的实用价值。此外,研究建立的耦合水库调度的分布式水文模型以及水库水文改变度计算方法,可为流域多水库联合生态调度方案的制订提供技术支撑。

关键词:水文改变度;水库调度;分布式水文模型;模拟;东江;径流;人类活动;水利工程

译文:

Study on Hydrological Regime Alteration of Lower Reaches of Dongjiang River Basin

Abstract: [Objective] The construction and operation of reservoirs will change the natural hydrological processes in the lower reaches of the river basin and cause ecological impacts. Quantitative evaluation of the impact of reservoirs on the hydrological process of the basin is the basis for improving water operations and restoring the downstream ecological flow process. Aiming at the problem of insufficient analysis of the influence of the degree of hydrological change on the reservoir operation in the current study, a new research idea is put forward. [Methods] Taking Dongjiang River Basin in Guangdong Province as an example, the distributed hydrological model coupled with reservoir dispatching process is adopted to calculate and analyze the hydrological process of Boluo Station in the lower reaches of the Dongjiang River after the construction of the reservoir and under different discharge discharge situations based on the current reservoir dispatching mode. To evaluate the influence of the construction and operation of the reservoir on the downstream hydrological regime under the same meteorological and underlying surface conditions. [Results] The result shows that the construction and operation of the three major reservoirs, Xinfengjiang, Fengshuba and Baipenzhu reservoirs in the Dongjiang River Basin, have significantly changed the monthly flow process, annual extreme flow, and annual runoff change rate of the downstream Boluo Station. In particular, the frequency of high and low flow rate, the rate of flow increase rate and decrease rate and the number of flow reversal have changed greatly. The overall hydrological change degree of Xinfengjiang single reservoir operation stage is 70.90%, the overall hydrological change degree of Fengshuba and Xinfengjiang double reservoir operation stage is 70.96%, and the overall hydrological change degree of the three reservoirs combined operation stage is 71.77%, all of which have been highly changed. The higher the monthly minimum discharge threshold of upstream reservoir, the more obvious the hydrological change degree of downstream. The hydrological change under Xinfengjiang reservior dam is the most significant. [Conclusion] The distributed hydrological model coupled with reservoir operation and the method of variation range of hydrological change index is an important means to evaluate the change of hydrological regime in a basin, and has great practical value for the study of hydrological regime in a basin that lacks measured runoff data before the construction of a reservoir. In addition, the distributed hydrological model of coupled reservoir scheduling and the calculation method of reservoir hydrological variation can provide technical support for the formulation of multi-reservoir joint ecological scheduling scheme.

Keywords: hydrological alternation degree; reservoir operation; distributed hydrological model; simulation; Dongjiang River; runoff; human activity; water conservancy project

案例五
原文：

In many societies a growing consensus has arisen about the importance of ecological knowledge: It helps us to address some of the most pressing problems we face at both global and regional level by providing research and effective management options to deal with global warming, diminishing natural resources and the deterioration of soils and water resources. Debates about natural disasters, about the purity of natural things, and about the perceived crisis of the nature-culture relationship in general all revolve around the role and the importance of ecological knowledge in social processes and in negotiations about the kind of nature with and in which we wish to live. For this reason it seems not only worthwhile but also essential, in a sense, to take a closer look at the logical and disciplinary construction of ecological knowledge from a philosophy of science perspective. Inversely, for the philosophy of science ecology is an interesting field having identified as one future perspective that it is important to link general philosophy of science with special philosophies of science in a more fruitful way. This is the framing I have in mind when I propose in the following to explore a distinct epistemological approach for ecological knowledge.

First of all, when "philosophy of science deals with the foundations and the methods of science" its attention is most often focused on the foundations and methods of a discipline. This already entails the expectation that scientific knowledge needs to be developed in a distinct institutional setting, in one that is in some sense a "closed society", with its own language and customs. This is certainly true of scientific disciplines such as physics or astronomy, which have an age-long tradition and a similarly long tradition of philosophical reflection. But there are other epistemic domains whose foundations and methods cannot be described adequately by framing them as a discipline in this sense. They should rather be conceived as a field of knowledge that is scattered between different academic disciplines and ultimately even beyond the academic context itself. I propose, first, that ecology be understood in the sense of a field of knowledge rather than as a discipline; and second, I propose an epistemological description to encompass this patchwork field. The theories in this field move back and forth between three basic conceptions and it is this productive movement, I argue, that stabilizes ecological knowledge.

At the end of this chapter will be offered another more general perspective that is a discussion on the relationship between the philosophy of nature and the philosophy of science in view of the field of ecological knowledge. I will provide some reflections on how the two might be linked and how this could help to better understand the plurality in the field.

译文：

　　在许多国家，越来越多人意识到了生态知识的重要性：我们面临着一些极为急迫的全球性或区域性问题,生态知识通过提供研究和有效管理方法帮助我们应对全球变暖、自然资源减少及土壤和水资源退化问题。关于自然灾害、天然物质的纯净度,以及我们已经意识到的自然与人类文化之间普遍存在的矛盾所引发的争论,都围绕着生态知识在社会发展进程中的角色,以及在我们生活共处于何种自然界的讨论中的重要性来进行。因此,从某种意义上说,从科学哲学的角度来进一步探讨生态知识的逻辑结构和学科建设不仅具有重要意义,也是十分必要的。反之,科学生态哲学是一个很有意思的领域,被认为是未来更为有效地结合通用科学哲学和专业科学哲学的角度。下文是我试着以认识论的角度来探讨生态知识的框架。

首先,"科学哲学处理的是科学的基础和方法",其关注的焦点是某一学科的基础和方法。这就已经说明了科学知识需要在一个有着自己的语言和习俗的、从某种意义上来说"封闭社会"的专业体制环境中发展。对于物理或天文学等一些有着漫长历史和同样漫长哲学反思历史的科学学科来说,的确如此。但从这个角度说,还有一些认知域的基础和方法不能简单归为一门学科,而应该被视为是一种知识领域,是不同学科的交叉,最终甚至超越了本身的学术语境。我建议,首先,生态学应被视为一种知识领域,而非一门学科;其次,我对这个跨学科领域提出了认知描述。这一领域的相关理论游走于三种基本概念之间。我认为,正是这一富有成果的交叉行为构建了生态知识。

本章结尾将提供另一种更为宏观的视角——以生态知识领域为背景,就自然哲学和科学哲学之间关系进行探讨。我将提供一些关于这两者结合方式,以及两者结合如何帮助我们更好地理解这一领域里两种哲学共存的一些想法。

第四节　翻译评析

工程专业学术研究论著翻译之"专业",在于译者须对论著内容的专业知识有一定的了解,具备足够的专业知识才能正确理解原文;翻译之"精准",在于准确、忠实地移植原文内涵与信息;翻译之"美学",在于译文重现原文严谨、客观的学术风格。

一、词法层面

工程专业学术研究论著多涉及术语。术语多是名词词组,多采用直译,必要时采用"增词"或"减词"等翻译技巧,补偿译语语言表述的缺省。

【例1】原文:按照有效应力原理,降雨引发的规模较小的浅层滑坡,主要受岩土体的内聚力控制,而规模较大的深层滑坡则主要受摩擦力控制。

译文:According to the principle of effective stress, for rainfall-induced landslides, the small and shallow landslides are mostly controlled by the cohesive strength; the large and deep landslides are controlled by friction.

分析:在该句中,"有效应力原理"和"内聚力"采用直译的翻译方法,处理方法较为简单,相应地译为"principle of effective stress"和"cohesive strength"。而"降雨型滑坡"则没有简单翻译为"rainfall-type landslides",而是根据语义结合其内涵——"降雨所导致的滑坡",增加"induced"一词,译为"rainfall-induced landslides"。在英译时,性质、类别或范畴类词语通常省略不译。"摩擦力"的"力"是术语中常见的定性分类型词语,此处翻译为"friction"。

【例2】原文:综合分析表明,所设计的7种调控方案均能有效地提升河网水动力,在开启西南侧河道的1台水泵和西部泵闸时,流场分布最均匀且水动力整体提升效果最佳。研究结果可为南方平原河网的水利工程水动力调控提供参考。

译文:The comprehensive analysis shows that the seven control schemes designed can effectively improve the hydrodynamic force of the river network. When one water pump and the west pump gate are opened in the southwest river channel, the flow field distribution is the most uniform and the overall hydrodynamic force improvement effect is the best. The research result can provide a reference for the hydrodynamic regulation of water conservancy projects in the southern plain river network.

分析:该案例中,"水动力"中的"水"没有简单译为"water",因其内涵与"水利"或"水电"相关,故译为"hydrodynamic force"。由此可见,词语的翻译应当根据上下文决定,准确翻译其内涵和意义。

【例3】原文:东江流域新丰江、枫树坝和白盆珠三大水库的修建和运行显著改变了下游博罗站的月流量过程、年极值流量以及年内径流改变率等水文指标,尤其是高低流量出现频率、流量增加率减少率和流量逆转次数改变较大。

译文：The result shows that the construction and operation of the three major reservoirs, Xinfengjiang, Fengshuba and Baipenzhu reservoirs in the Dongjiang River Basin, have significantly changed the monthly flow process, annual extreme flow, and annual runoff change rate of the downstream Boluo Station. In particular, the frequency of high and low flow rate, the rate of flow increase rate and decrease rate and the number of flow reversal have changed greatly.

分析：该案例中出现了"月流量过程""年极值流量""年内径流改变率"等术语。这些术语内在逻辑关系较为简单，词义清晰明确，因此采用直译的处理方式，分别译为"monthly flow process""annual extreme flow"和"annual runoff change rate"。

二、句法层面

中文中存在大量的无主句。无主句通过省略不重要或不必要的、意义不言自明的成分，形成结构简洁的句子，重点信息更为突出。

【例4】原文：详细研究了三峡地区部分县市的滑坡和降雨历史资料，从滑坡与降雨量、暴雨以及降雨时间3个不同角度的关系分析了降雨与降雨型滑坡的关系。

译文：Historical data of landslides and rainfall of some counties and cities in the Three Gorges area are investigated in detail, and the relationship between rainfall and rainfall-induced landslides is analyzed from three different points of view, rainfall, rainstorm and rainfall duration.

【例5】原文：在此基础上，提出了降雨因子的概念。同时，还提出了一种预报降雨型滑坡的新方法，定量化地描述了降雨型滑坡的易发程度。按照一定的标准，对每种降雨分因子进行分级，通过多因子叠合分析来研究降雨因子与降雨型滑坡之间的关系，并据此准确地预报滑坡的易发程度。

译文：The conception of rainfall-factor is suggested thereafter. At the same time, a new method of predicting rainfall-induced landslides is proposed to depict the probability of rainfall-induced landslides quantitatively. Every rainfall-sub-factor is graded according to some criteria, and then the relationship between rainfall-factor and rainfall-induced landslides is studied.

分析：在以上两个案例中，原文出现了大量无主语句子，如"详细研究了……分析了降雨与降雨型滑坡的关系""提出了……。同时，还提出了……"。中文中使用无主语句子，通常因为主语不重要或泛指。但在英语中，除祈使句等特殊句式外，主语必不可少。因而中文的无主句，通常翻译为英文的被动句。被动句天生具有语气客观、结构简洁的特点。

【例6】原文：通过将这种滑坡预报新方法应用于三峡的万县地区，证明了它可以比较准确地确定滑坡发生的时间。这种滑坡预报方法将为根据历史降雨和滑坡资料来预测降雨型滑坡奠定良好基础。

译文：It is proved that this new method of landslide prediction can predict the occurrence time of landslides with higher precision through applying it to Wanxian, a city in the Three Gorges area. This method of landslides prediction would lay a solid foundation for predicting rainfall-induced landslide according to historical data of rainfall and landslides.

分析：在该案例中，原文的无主句"证明了……"没有翻译为常见的被动句，而是采用"It is + 过去分词 + that 从句"的形式。It 作为形式主语的特殊句式同样省略了真正的主语，表述地道，语言精练。

三、篇章层面

【例7】原文：In many societies a growing consensus has arisen about the importance of ecological knowledge: It helps us to address some of the most pressing problems we face at both global and regional level by providing research and effective management options to deal with global warming, diminishing natural resources and the deterioration of soils and water resources. Debates about natural disasters, about the purity of natural things, and about the perceived crisis of the nature-culture relationship in general all revolve around the role and the importance of ecological knowledge in social processes and in negotiations about the kind of nature with and in which we wish to live. For this reason it seems not only worthwhile but also essential, in a sense, to take a closer look at the logical and disciplinary construction of ecological knowledge from a philosophy of science perspective. Inversely, for the philosophy of science ecology is an interesting field having identified as one future perspective that it is important to link general philosophy of science with special philosophies of science in a more fruitful way. This is the framing I have in mind when I propose in the following to explore a distinct epistemological approach for ecological knowledge.

译文:在许多国家,越来越多人意识到了生态知识的重要性——我们面临着一些极为急迫的全球性或区域性问题,生态知识通过提供研究和有效管理方法帮助我们应对全球变暖、自然资源减少及土壤和水资源退化问题。关于自然灾害、天然物质的纯净度,以及我们已经意识到的自然与人类文化之间普遍存在的矛盾所引发的争论,都围绕着生态知识在社会发展进程中的角色,以及在我们共处于何种自然界的讨论中的重要性来进行。因此,从某种意义上说,从科学哲学的角度来进一步探讨生态知识的逻辑结构和学科建设不仅具有重要意义,也是十分必要的。反之,科学生态哲学是一个很有意思的领域,被认为是未来更为有效地结合通用科学哲学和专业科学哲学的角度。下文是我试着以认识论的角度来探讨生态知识的框架。

分析:在该案例中,原文采用了"it""this""for this reason""inversely"等代词、介词短语、副词以及各类从句表达上下文的逻辑关系,在中文译文中,代词的处理方式通常为"还原",即将代词还原为其所指代的名词或名词词组。英文通常采用某个词语表达逻辑关系;而中文则相反,多省略逻辑关系词,通过语义组合,让读者推断上下文的逻辑关系。因此,翻译时应注意顺应译语的表达习惯。

第五节　拓展延伸

一、中译英

水利枢纽调压阀仿真及试验研究

摘　要:以某水利枢纽为研究背景,针对其调压阀进行仿真模拟研究,分析不同工况下,水利枢纽的流量特性及气蚀情况。结果表明,开度较小时,水利枢纽调压阀的流阻系数变化趋势显著,下降趋势接近线性减小趋势;随着开度的增大,水利枢纽调压阀流阻系数的变化趋势逐渐趋于平缓。表明当开度较小时,开度的变化对于水利枢纽调压阀流阻系数的变化影响较大。最低压力对水利枢纽调压阀气蚀现象的影响较小,决定水利枢纽调压阀是否发生气蚀现象的主要因素在于其流阻系数的大小。上述结果显示,仿真模拟的准确性较高,可采用仿真模拟对水利枢纽调压阀的流量特性及气蚀情况进行分析。

关键词:水利枢纽;流量特性;开度;气蚀

二、英译中

The present approach argues that the existence of a plurality of theories (or programmes) can have a positive impact because it allows for greater logical flexibility and thus for more explanatory power as well. On this point, Nancy Cartwright (1999) has shown that, if anything, the search for explanatory unity detracts from the search for truth. In a similar vein, Shrader-Frechette and McCoy (1994) argue in the course of their attempt to develop a philosophical vocabulary adequate to partial knowledge that, in ecology, the main method used for linking data with a hypothesis is not the classical deductive scheme but rather (in case studies, for instance) a variety of logic they call "informal inferences". Cooper (1998) suggests a three-fold scheme in which theoretical principles, phenomenological patterns and causal generalizations are the basic forms of generalization in ecology. This constitutes a philosophical taxonomy which, as Cooper points out, should not be taken as a rigorous or categorical classification; instead, it should function as an aid to distinguishing the different modes of investigation (model-driven or data-driven ecology, for example) and acknowledging their varied generalizations while not dismissing the possibility that laws may exist in ecology.

The various regions in the taxonomic space. . . are all more or less occupied. There is a great deal of variation in scope and reliability among ecological generalizations. . . laws are what philosophy of science has tended to take them to be, then there are no laws in biology. But that does not mean that everything in biology is equally contingent. (Cooper, 1998, p. 582, 584)

Thus all these authors share an unease toward any logical or methodological unity in science. Instead, they support the idea of different epistemological strategies that can be described in a philosophically sound way. In ecology these epistemological strategies range from experimental studies in the lab, through real-world experiments, quasi-experimental studies and case studies, to purely observational studies in the field.

参 考 文 献

［1］Patrind hydropower project［EB/OL］. https://www. adb. org/sites/default/ files/project-documents/44914/44914-014-esmr-en_2. pdf.

［2］戴光荣,王华树. 翻译技术实践教程［M］. 北京:北京大学出版社,2022.

［3］单宇,范武邱,谢菲. 国内科技翻译研究(1985—2015)可视化分析［J］. 上海翻译,2017(2):34 – 42.

［4］丁继新,尚彦军,杨志法,等,降雨型滑坡预报新方法［J］. 岩石力学与工程学报. 2004,23(21):3738 – 3743.

［5］法律出版社法规中心. 中华人民共和国建筑法:注释本［M］. 北京:法律出版社,2021.

［6］中国国家标准化管理委员会. 翻译服务规范　第 1 部分:笔译:GB/T 19363. 1—2008［S］. 北京:中国标准出版社,2008.

［7］中国国家标准化管理委员会. 翻译服务译文质量要求:GB/T 19682— 2005［S］. 北京:中国标准出版社,2005.

［8］范英琦,郑航,刘悦忆,等. 东江流域下游水文情势的改变度研究［J］. 水利水电技术(中英文),2023(7):77 – 87.

［9］方梦之,范武邱. 科技翻译教程:第二版［M］. 上海:上海外语教育出版社,2015.

［10］Jiangbo. 江西省峡江水利枢纽工程简介［EB/OL］. (2019 – 09 – 25)［2023 – 08 – 28］. http://www. jxxjslsn. com/gcgl/sngc/2019/09/24135. shtml.

［11］金海. 水利工程技术标准术语和常用词翻译的一致性管理［J］. 中国科技翻译,2015,28(4):12 – 15.

［12］李长栓. 非文学翻译理论与实践:理解、表达、变通［M］. 北京:中译出版社,2022.

［13］刘峥,张峰. 对科技翻译研究困境的再思考［J］. 中国科技翻译,2014

（2）:47 - 49,59.

[14]麻土华,郑爱平,李长江.降雨型滑坡的机理及其启示[J].科技通报.
2014(1):39 - 43,71.

[15]全国人大常委会法制工作委员会.中华人民共和国民法典:汉英双语
版[M].北京:法律出版社,2021.

[16]任文,李长栓.理解当代中国:高级汉英笔译教程[M].北京:外语教学
与研究出版社,2022.

[17]史澎海.工程英语翻译[M].西安:陕西师范大学出版社,2011.

[18]孙建光,李梓.工程技术英语翻译教程[M].南京:南京大学出版社,
2021.

[19]孙万彪.英汉法律翻译教程[M].上海:上海外语教育出版社,2003.

[20]涂成颉.水利枢纽调压阀仿真及试验研究[J].水利科技与经济,2023
(6):64 - 68.

[21]王华树.翻译技术研究[M].北京:外语教学与研究出版社,2023.

[22]谢龙水.工程技术英语翻译导论[M].北京:北京希望电子出版社,
2015.

[23]徐存东,訾亚辉,黄嵩,等.基于 MIKE 21 的圩区河网水动力调控方法
研究[J].水利水电技术（中英文）,2023(7):161 - 170.

[24]许建平.英汉互译入门教程:第二版[M].北京:清华大学出版社,
2015.

[25]闫文培.实用科技英语翻译要义[M].北京:科学出版社,2008.

[26]岳峰,曾水波.科技翻译教程[M].北京:北京大学出版社,2022.

[27]张铎,李仙仙.国际工程项目翻译管理工作问题及对策[J].水利水电
工程设计,2019(4):51 - 53.

[28]张法连,崔璨.知识翻译学视域下的法律翻译[J].当代外语研究,2023
(6):25 - 32.

[29]中国电建集团成都勘测设计研究院有限公司.水利水电工程英汉图文
辞典:地质卷[M].北京:中国水利水电出版社,2022.

[30]中国工程咨询协会.施工合同条件[M].北京:机械工业出版社,2002.

［31］中国翻译协会. 中国语言服务行业规范之一：本地化业务基本术语［EB/OL］. (2016－01－12)［2023－08－11］. http://www. tac-online. org. cn/index. php? m＝content&c＝index&a＝show&catid＝389&id＝1238.

［31］中国翻译协会. 中国翻译人才发展报告［R］. 2022.

［32］中国翻译协会. 中国翻译及语言服务行业发展报告［R］. 2022.